# Mushrooms
## and Toadstools

OF BRITAIN AND EUROPE

GREEN GUIDE

# Mushrooms and Toadstools

## OF BRITAIN AND EUROPE


**Gordon Dickson**

Illustrated by Catharine Slade

B L O O M S B U R Y

LONDON • OXFORD • NEW YORK • NEW DELHI • SYDNEY

Bloomsbury Natural History
An imprint of Bloomsbury Publishing Plc

50 Bedford Square
London
WC1B 3DP
UK

1385 Broadway
New York
NY 10018
USA

www.bloomsbury.com



First published by New Holland (UK) Ltd, 1990
This edition first published by Bloomsbury 2016



British Library Cataloguing-in-Publication Data
A catalogue record for this book is available from the British Library.

Library of Congress Cataloguing-in-Publication data has been applied for.

ISBN: PB: 978-1-4729-2714-9
ePDF: 978-1-4729-2716-3
ePub: 978-1-4729-2715-6

2 4 6 8 10 9 7 5 3 1

Designed and typeset in UK by Susan McIntyre
Printed and bound in China by C&C Offset Printing Co., Ltd

Frontispiece: *Coprinus comatus*, p.74

# Contents

# Introduction

There are several thousand species of fungi to be found throughout Europe, including the British Isles. Many are microscopic and, though they are important to the agriculturalist because of the damage they cause to crops, they are not considered in this guide. Of the larger fungi there are at least 2,000 species, of which nearly 1,000 are common *somewhere*, but most of those common in the north are rare in the south and vice versa.

The species included in this book are mainly those that are recognised readily in the field without the need for a microscope (though a simple hand lens is always a help) as they have some distinguishing character of appearance, smell or taste.

They are arranged in groups, starting with the fungi that have tubes, followed by those with gills. Next come the stomach-fungi, most of which look like round balls, then the jelly-fungi and the brackets that grow on trees (though if a bracket has gills it is shown with the other gill-fungi). Last are the disc-fungi and the black Pyrenomycetes.

Size varies considerably within species and individuals will almost always be found that are outside the dimensions given. The thickness of the flesh depends on the species and also on the size of the cap. The cultivated mushroom, for example, is referred to as 'thick-fleshed', while the little Mycenas, so common on the woodland floor, have very little material between the cap surface and the gills and are referred to as 'thin-fleshed'.

Although some fungi are described as associating with specific trees there are exceptions when the host tree may be of a different species from that stated. However, there are very few species of fungi that will grow on both deciduous and coniferous trees.

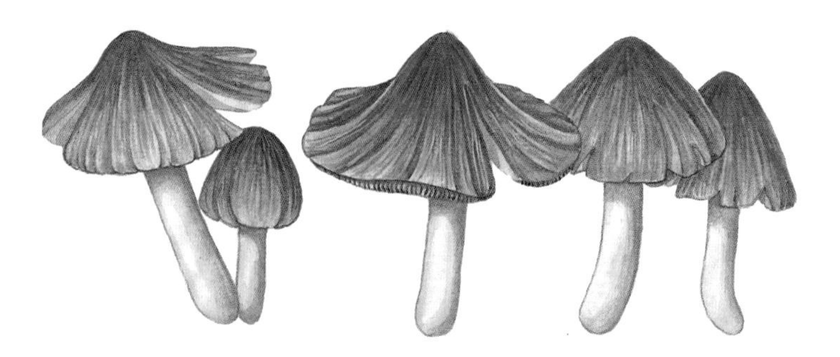

*Macrolepiota procera*, p.46

A far greater number of fungi fruit in the autumn than in the spring. Unless otherwise stated, the species described may be found between May and October in northern areas of Europe and a month later in southern areas – in most years the peak months are August to September and September to October respectively. Drought followed by rain usually produces extensive fruiting of Agarics and sharp frost puts an end to it, though a mild winter may allow fruiting to continue right through into spring.

# What are fungi?

For centuries it has been customary to classify the living world into two main kingdoms – animal and vegetable – and, until comparatively recent times, it seemed obvious to include fungi in the latter category. However, lacking the pigment known as chlorophyll, which is contained in green plants, fungi are unable to make use of the sun's energy to build up the carbohydrates from which a plant is formed. Instead, like animals, they obtain their energy through the breakdown of organic matter. Further, the hard parts of fungi are chemically closer to the chitin that forms the casing of insects than to lignin, which forms the hard parts of plants. A third category – the fungal kingdom – has had to be created to house these extraordinary organisms.

A mushroom or toadstool is only the fruitbody, or reproductive part, of an extensive network of very fine threads that branch, join and weave below

▲ *Boletus chrysenteron*, p.24

the surface of the ground, breaking down decaying material for sustenance. In the same way, the bracket fungus that grows on a tree is the fruit of a body of fine threads that penetrate the substance of the wood. This network is known as the mycelium and is common to all fungi. The individual threads (hyphae) are too small to be seen with the naked eye, but often a number of hyphae will cluster together to form visible threads about the thickness of sewing cotton. Certain fungi form even thicker black threads like bootlaces, which are known as rhizomorphs, and these may be seen under the loose bark of dead trees.

The mycelium is perennial and certainly persists in the soil for decades, probably for centuries and possibly for millennia. It plays a major part in the rotting down of the leaves that fall in autumn – which would be metres deep in a very short time without this process of decomposition.

Not all fungi limit themselves to the breakdown of leaves, however; there are many that send their hyphae down to the roots of trees and form a network within the outer layers of the finer roots. There, a mutual interchange of material, known as mycorrhizal symbiosis, takes place to the benefit of both fungus and tree, and without which the tree would not prosper. It is believed that the fungal hyphae are able to take up and transfer to the tree minerals such as phosphates, and to receive from the tree compounds that they are themselves unable to synthesise. Foresters now regard the fungal element in the soil as extremely important.

# How to identify fungi

Nature seems to have explored every avenue of shape, texture and colour in fungal fruitbodies, from cups to finger-like projections and from caps on stalks to irregular potato-like lumps buried in the ground, from soft jellies to brackets too hard to be cut with a knife and from brilliant red to deep turquoise. They fall into a few broad divisions that help to classify the species, and these are explained on pages 13 to 19.

The typical mushroom fruitbody consists of a cap, which is generally round, and a stem, which is usually attached to the centre of the cap but may also be attached at the edge.

Beneath the cap are the spore-bearing surfaces, which vary according to the division to which the fungus belongs and may be either flat vertical plates known as gills (in the Agaricales) or tubes (in the Boletales) which end in openings known as pores. It is important to note the arrangement of the gills, as these are often characteristic of a particular genus; they may run down the stem (decurrent), be free of the stem (free), or just touch it (adnate).

The stem itself may have a ring, which can be either firm or fragile. This is formed from a membrane (the partial veil) that protects the developing gills of the young mushroom and breaks away from the edge of the cap as it opens. The base of the stem may be contained in a bag, known as the

▼ *Polyporus squamosus*, p.90

volva. This is the remains of the sheath (the universal veil) that completely surrounds both cap and stem as the fruitbody grows, rupturing as the mushroom rises from the ground. Both veils may or may not be present and it is extremely important to recognise them if the fungus is to be identified correctly, particularly as the volva is a characteristic of some of the most deadly poisonous fungi. A penknife or garden trowel can be used to lift the stem base out of the ground for examination.

Spore colour can also be an aid to identification but whilst the shade can usually be assessed from the colour of the gills, this is no more than a rough guide. A coating of spores on adjacent caps or on leaves close to the fungus may provide further evidence but an accurate assessment of colour can be made only by taking a sample. The stem should be cut off the specimen and the cap, moistened on the upper surface, laid gills down on a sheet of white paper and covered with a tumbler. After a few hours a thick dusting of spores should have formed on the paper – unless the inevitable maggots have reduced everything to a pulp.

It is helpful when identifying fungi to note which trees are in the vicinity as some are quite specific in their mycorrhizal associates. Additional information may be gathered by ascertaining whether a fungus is growing on the ground or on wood, which may not be very obvious if the wood is buried.

▼ *Collybia dryophila*, p.50

▲ *Stropharia aeruginosa*, p.71

# Poisonous fungi ☠

The main poisonous species are identified in this book, even if they are not common. The edible fungi included are those that are unlikely to be confused with harmful species of similar appearance. Tasting as a diagnostic method should be employed only on those types where it is indicated and the remnants should not be swallowed.

It is vital that anyone collecting mushrooms is able to identify positively those species that are edible; if there is any doubt about a particular specimen, an expert opinion should be sought. Serious cases of poisoning have occurred through the failure to recognise *all* the distinguishing features of a species.

# Conservation of fungi

In continental Europe, especially in the east, there has been a worrying decline in the number of fungus fruitbodies found in recent years. This has been particularly noticed in the edible species that have been marketed. There is also some evidence to indicate that the disappearance of the mycorrhizal species has preceded the wholesale death of forests that has occurred in eastern Europe. In the Netherlands a severe decline in the number of Chanterelles collected has been noted. It seems likely that this has been caused by environmental factors such as acid rain, heavy metals, nitrogenous compounds and so on, rather than by collecting. It would seem unlikely that, in the short term, collecting of fruitbodies from perennial mycelia would produce any dramatic change.

In Britain there has in the recent past been little, if any, commercial collecting and no recent declines that could be attributed to it have taken place. Small amounts of collecting 'for the pot' are therefore permissible at present. Reduction in numbers of British species as has been noted has been due to destruction of habitat by the application of inorganic fertilisers to pasture land, conversion of heathland and deciduous woodland into conifer plantations, and the removal of dead wood for fuel. Species that grow on herbivore dung show a dramatic decline when artificial fertilisers are applied to grassland and the relatively small areas of unimproved pasture require urgent conservation.

# Nomenclature

All fungi have scientific names that comprise two parts: the first part identifies the genus to which the species belongs; the second part which may have been created in acknowledgement of an individual or to serve as a description of certain features, identifies the species within the genus.

▼ *Pluteus cervinus*, p.61

Few fungi have common names that are in general use, and only those that are recognised as valid are given in this book. Translations of the second part of scientific names are provided as they often serve as a helpful reminder of why certain species were so labelled. In some cases, however, the intentions of the originator remain obscure.

# Classes of fungi

Whilst all fungi have the same basic structure of hyphae and reproduce by means of spores, the way in which they distribute their spores separates them into two main divisions: Basidiomycetes and Ascomycetes.

Basidiomycetes allow their spores to fall into the passing air currents and so they have to elevate their fruitbodies at least a short way above the ground. The spores are formed on the stalks of cells known as basidia, and there are usually four to a cell.

Ascomycetes form their spores in tubes like gun barrels known as asci, which usually point upwards. When ripe, the spores are forcibly ejected for several millimetres – which is enough to get them airborne. In some species, small changes in humidity or air pressure are enough to initiate this release and, as many asci discharge at the same time, the spores appear as a little puff of smoke.

▼ *Clitocybe nebularis*, p.55

▲ *Agaricus campestris*, p.69

Perhaps fortunately, very few fungal spores ever germinate, or at least few develop into further organisms, and for this reason they are produced in enormous numbers, an average mushroom producing from 10 to 20 thousand million. *Langermannia gigantea*, the Giant Puffball, has been calculated to produce over 10 billion spores from one average fruitbody. It seems that the likelihood of them falling on suitable ground that is not already inhabited by an antagonistic fungus is extremely small.

## BASIDIOMYCETES

### Boletales

With the exception of two rare species, all European *Boleti* grow on the ground. All form mycorrhizal associations with trees (or, very occasionally, with flowering plants). They have the typical 'toadstool' shape and on the undersurface of the cap they have tubes instead of gills. These tubes, when viewed from the surface, have the appearance of a sponge but if they are cut vertically the tubular structure, aligned to allow the spores to fall freely, can be seen. The flesh is usually white or cream in the cap but, if cut, may change dramatically within seconds to some shade of blue or green, and the entire fungus may discolour rapidly on handling. Genera in this group include:

*Boletus*, which has a dry cap and a stem that may be covered with small dots or a fine net;

*Suillus*, characterised by a cap that is slimy in wet weather and a smooth stem;

*Crepidotus mollis*, p.37

*Leccinum*, which has a dry cap and is characterised by a scaly stem;

*Tylopilus*, which has pink pores and a dark net on the stem.

## Agaricales

These, like the Boletales, have a typical 'toadstool' shape but have gills, not tubes, on the undersurface of the cap. The spores are formed on the surfaces of the gills, from whence they fall into the airstream for dispersal. A few Agaricales have lateral stems but in most cases the cap is raised from the ground by a vertical stem of varying length.

The genera *Lactarius* and *Russula* have, for the purposes of this book where identification is based on simple macroscopic characters, been grouped with the Agaricales in view of their gross similarities. However, they have a different microscopic structure from other gill-fungi and, in fact, form a separate group, the Russulales. They are described below, together with other important genera.

*Lactarius* (Milk-caps): It is characteristic of this genus that, when the cap or gills are broken, a milky juice quickly oozes out of the damaged surface. In some species, the colour of the milk changes after a few minutes. It is non-toxic but can have an extremely hot taste. The gills are thick, adnate or decurrent, never free, and the spores are white or cream. *Lactarius* species may vary considerably in size but they are usually fairly sturdy fungi.

▲ *Hygrophoropsis aurantiaca*, p.57

***Russula:*** These are hard, brittle-fleshed fungi with white to yellow gills, most of which run all the way from the stem to the edge of the cap. Characteristically, the caps are rounded and fairly thick-fleshed, and in many cases brightly coloured. The stems are medium to tall, parallel and usually white, though some may have varying degrees of pink coloration. The spores range from white to deep saffron in colour.

***Hygrophorus*** and ***Hygrocybe:*** Together these genera form a group of fungi called 'Wax-caps', an apt term as the gills are thick, well spaced, and waxy-looking. They were at one time combined in the genus *Hygrophorus*, which is now limited to the pale-coloured species which occur in woodland. Those that grow in grassland are placed in two genera, *Camarophyllus* and *Hygrocybe*. Many are among the most brightly coloured fungi to be found. They vary in size from small to medium and in shape from a rounded cone, which never opens fully, to a flat-topped inverted cone. In moist weather the cap is slippery.

***Amanita:*** The genus contains some of the most poisonous fungi. It is quite a large group and only the most common and the most dangerous species are described in this book. Its most characteristic feature is the possession of a volva; this can be large and tough, or fragile, or reduced to a series of rings on the lower part of the stem. In no other genus is it so important to dig up the entire base when collecting. Most *Amanitas* have a ring as well as a volva but in the case of the Grisettes the ring is hidden at the base of the stem.

Aleuria aurantia, p.93

However, as rings can be lost it would be unwise to rely on this character as a test of edibility.

*Lepiota:* All are white or whitish fungi with white spores, a more or less scaly cap, white gills, which are free of the stem, and a ring. The genus is now divided into the large (edible) species, which are called *Macrolepiota*, and the small (inedible) species, which retain the name *Lepiota*, once used for all the species.

*Collybia:* This is a genus of fungi responsible for the breakdown of wood, leaf and needle litter in the woodland. The white gills are either free of, or lightly attached to the stem, which is characteristically tough, giving the whole group the name of 'Tough-shanks'. There is no ring.

*Clitocybe:* This genus is characterised by having white gills that are markedly decurrent. The cap is often funnel-shaped and the stem fairly short. They grow in deciduous woodland.

*Mycena:* These are small, white-gilled fungi, the gills mostly adnate, usually with conical caps which may later open flat. The stems are long and slim, and some species have clear or coloured sap, which oozes out of the stem when it is cut. They may grow on leaves, twigs, dead wood or moss, and may be solitary or tufted.

▲ *Pholiota squarrosa*, p.63

*Coprinus:* Many members of this genus spread their spores in an unusual way: the gills and cap slowly liquefy and the spores are washed away by rain, instead of becoming airborne. Most of the species that undergo this change live on buried wood so the spores are more likely to find satisfactory substrata than they would if dispersed by air. *Coprinus* species are usually either white or shades of brown and are proportionally tall for their width.

**Gastromycetes**

Known as the 'stomach-fungi', the members of this very mixed group form their spores on a network of hyphae and basidia, which is initially contained in the fruitbody. These may be expelled through a central opening – as in the Puffballs – or released by disintegration of the casing. In the Stinkhorns, a stem forms, on top of which is a slime containing the spores; this mucus is removed by flies, which thus distribute the spores. Some of the Truffles belong to this group. They are formed underground and give off a scent that makes them attractive to some animals as food; when eaten, the spores remain resistant to digestion and are voided with the animals' excreta and dispersed.

**Heterobasidiomycetes**

This is a group known as 'jelly-fungi' because of the elastic nature of the fruitbody. The spores are formed internally but are borne on basidia. The species described are all parasitic or saprophytic on wood. The genus *Tremella* is characteristic of the group.

▲ *Phlebia radiata*, p.83

### Aphyllophorales

These are fungi without gills. Perhaps the most illustrative is the genus *Stereum*, which consists of horizontally disposed plates with the spore-bearing surface on the underside.

## ASCOMYCETES

The larger Ascomycetes can be divided into two groups.

### Discomycetes

Known as the disc-fungi, these are characteristically saucer-shaped and of various sizes from less than a millimetre to several centimetres in width. The spore-producing layer is borne on the upper surface. Some Discomycetes, among which are the Morels, have a stem carrying a convoluted head.

### Pyrenomycetes

These generally have a black, crusty appearance that gives them the name of 'burnt' fungi. They bear their spores in asci, which point inwards to a flask-shaped hollow and make their way to the air through an opening in the flask. In addition, some Pyrenomycetes produce conidiospores, which are in effect the budding off of the ends of the hyphae; they may form a white dust on the surface of the fruitbody.

▲ *Amanita pantherina*, p.44

▼ *Paxillus involutus*, p.66

# Glossary

**Adnate** (gills) Attached to the stem

**Agaric** Fungus with gills

**Ascus** Microscopic part of an Ascomycete which fires the spores

**Basidium** Microscopic spore-bearing part of a Basidiomycete

**Cap** Upper part of a mushroom

**Decurrent** (gills) Running down the stem

**Free** (gills) Not touching the stem

**Fruitbody** Reproductive part of the fungus

**Gills** Flat, spore-bearing plates beneath the cap

**Gluten** Slimy coating of some fungus bodies

**Hyphae** Fine threads of which the entire fungus is composed

**Mycelium** Network of threads (hyphae), which form the unseen part of the fungus

**Partial veil** Membrane that protects the developing gills

**Pores** Openings of the spore-bearing tubes in Boletes

**Rhizomorphs** Thick, black threads formed from hyphae

**Ring** Formative remnant of hyphae surrounding the stem

**Spore** Single-celled equivalent of a seed in flowering plants

**Tubes** Spore-bearing structures (in Boletes) beneath the cap

**Universal veil** Sheath that surrounds the cap and stem of the developing fungus

**Volva** Remains of the universal veil, forming a bag at the base of the stem

▲ *Oudemansiella mucida*, p.48

# The Fungi

## *Boletus chrysenteron* POISONOUS ☠

Medium sized. Cap averages 6cm across; shades of hazel-brown; initially has slightly velvety texture that soon wears off, leaving it smooth and cracking. Pores dirty yellow. Flesh medium thickness, cream or straw, with thin reddish layer immediately under the cuticle; often turns pink where it has been damaged by slugs. Stem usually flushed red throughout most of its length except for the very top, which is yellowish. Found in deciduous woodland. Common. Common name: Red-cracking Bolete, though it rarely cracks red. Specific name, *chrysenteron*, means 'golden gut'.

## *Boletus subtomentosus*

Medium size, similar to *Boletus chrysenteron*. Cap averages 6cm across; mid to cigar-brown, covered initially with yellowish velvety down that wears off, allowing yellow cracks to appear. Pores dirty lemon-yellow, blueing slightly on bruising. Flesh creamy to yellow, not changing when damaged. Stem mainly brown with brick-red base but creamy in upper part. Found in deciduous woodland. Common. Specific name, *subtomentosus*, means 'slightly downy'.

## *Boletus edulis*

Large and thick-fleshed. Cap averages 12cm across; brown, rounded when young, flattened later. Pores white, becoming creamy-grey then yellow. Flesh white, pleasant smelling. Stem pale grey with white net, especially at the top; fat and bulbous. Edible, one of the best species. Found in woodland of all types, especially beech and oak. Common names: Penny Bun; Cep. Specific name, *edulis*, means 'eatable'. Closely related species with very pale pores include *B. aereus* with darker stem, net brown at apex, and *B. aestivalis*, cap and stem snuff-brown. Both edible.

## *Tylopilus felleus*

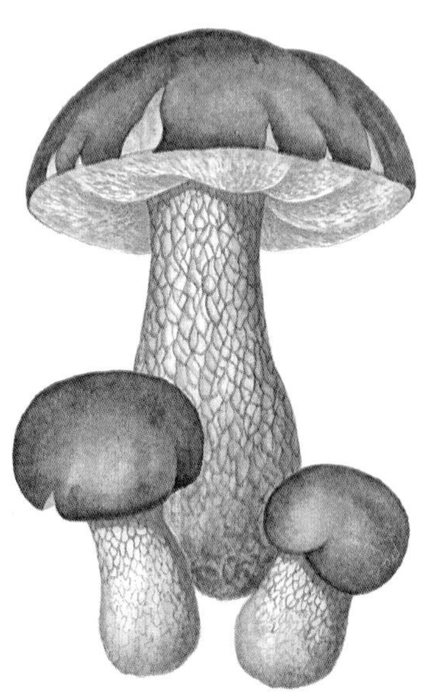

Fairly large. Cap up to 12cm across; snuff-brown; rounded at first but often becoming inverted so that the pinkish tubes are visible. Flesh white or cream, later becoming pinky-buff. Stem thick, pale at first but darkening to approach cap colour and covered in brown net. Tastes unpleasantly bitter but is not poisonous. Found mainly in deciduous woodlands. Commoner on basic soils. Common name: Bitter Bolete. Specific name, *felleus*, means 'like gall'.

## *Suillus bovinus*

Fairly large. Cap averages 8cm across; pale cinnamon with persistently paler margin; covered with slippery gluten. Pores pale grey to cinnamon; large, irregular with smaller ones inside, and slightly decurrent. Flesh slightly yellowish becoming clay-pink on exposure, darker in stem. Stem parallel, similar in colour to cap and usually rather short. Edible, but gluten best removed first. Common in Scots pine woods. Specific name, *bovinus*, means 'pertaining to oxen'.

## *Suillus luteus*

Fairly large. Cap averages 8cm across; chestnut coloured, paling to sepia at edge; covered with slippery gluten in wet weather. Pores and tubes dirty yellowish, small. Flesh pale, almost white, greying in stem base. Stem whitish at base, yellower nearer the top with a ring, initially pale but darkening almost to cap colour. Edible and tasty when gluten removed. Fairly common in Scots Pine woods. Common name: Slippery Jack. Specific name, *luteus*, means 'yellow'. Similar *S. grevillei* is paler and grows with larches.

## *Leccinum scabrum*

Usually large, but varies greatly in size. Cap snuff-brown, rounded; feels as if it is filled with cotton wool. Pores off-white. Flesh white, unchanging when cut. Stem tall, white, covered with blackish scales. Found with birch. Common name: Birch Bolete. Specific name, *scabrum*, means 'rough'. *L variicolor* is a slightly darker fungus with stem flesh turning blue-green after cutting.

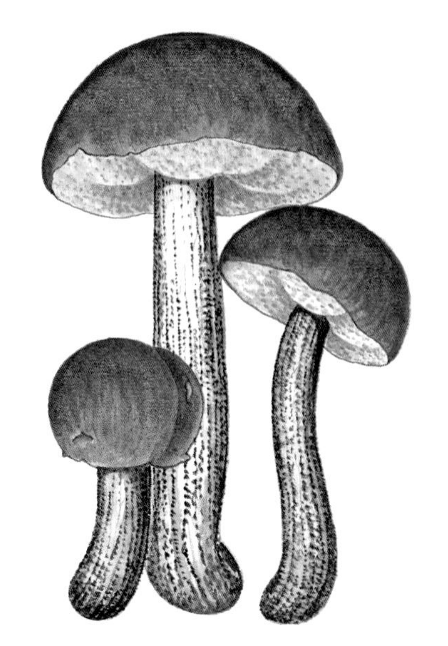

## *Leccinum versipelle*

Large fruitbody. Cap averages 12cm across; orange; dry and much firmer to the touch than *L. scabrum*; the cap cuticle often forms an overhanging skirt. Pores small, greyish. Flesh white, rapidly darkening to blue-green on cutting, soon becoming nearly black in both stem and cap. Stem white covered with brownish-black scales, bruising black. Found with birches. Edible. Common name: Orange Bolete. Specific name, *versipelle*, means 'of changeable appearance'.

### *Lactarius torminosus* **POISONOUS** ☠

Medium-sized fungus. Cap averages 7cm across; orange, zoned, covered with an orange-red coat like matted wool. Gills white, slightly decurrent. Flesh fairly thick. Milk white, hot taste. Stem short to medium, sturdy; pale flesh-colour without hairs. Poisonous but not deadly. Grows with birch. Common name: Woolly Milk-cap. Specific name, *torminosus*, means 'griping'. A very similar but much paler species is *L pubescens*, which is also poisonous.

### *Lactarius deliciosus*

Medium sized. Cap averages 7cm across; zoned orange with touches of green. Gills pale pink, greening with age or damage; decurrent. Flesh pale yellowish when first cut. Milk carrot colour, not changing; taste mild or very slightly bitter. Stem short, stout; greyish-buff, usually with depressed spots coloured orange. Edible. Grows in woodland, usually with pines. Found throughout Britain but rather more frequent further north. Common name: Saffron Milk-cap. Specific name, *deliciosus*, means 'delicious'. *L. deterrimus* is similar; milk turns purplish in ten minutes and wine-red in half an hour. Also edible but not as highly esteemed.

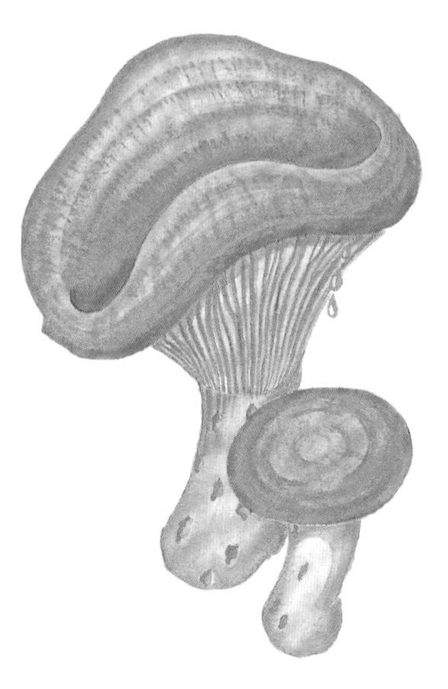

## *Lactarius blennius*

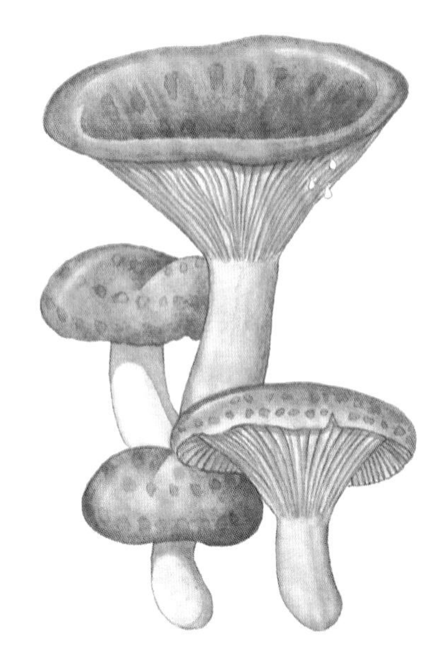

Medium size. Cap averages 6cm across; various shades of brown and greenish-grey, often with blotches of darker colour in concentric zones; rounded at first, becoming funnel-shaped; slimy. Gills white becoming grey; slightly decurrent. Flesh whitish. Milk white at first, turning grey on the gill; it is intensely hot and acrid. Stem short and sturdy, greyish-cream. Grows with beech and oak. Common. Specific name, *blennius*, means 'slimy'.

## *Lactarius turpis*

Large. Cap averages 15cm; dark olive-brown to dark grey; thick, with central depression when old. Gills creamy, decurrent. Flesh white, browning when cut. Milk white; taste very hot. Stem paler, stout and short so cap lies almost on the ground. The cap contains a chemical indicator which will react if a few drops of household ammonia are placed on the surface, turning the area bright purple. Grows with birch. Common name: Ugly Milk-cap. Specific name, *turpis*, means 'ugly'.

## *Lactarius rufus*

Medium to large. Cap averages 8cm, reddish-brown to rusty; dry, almost always with a central umbo. Gills white, tinged with cap colour; slightly decurrent. Flesh white. Milk white; tasting mild at first, then, after a delay of about a minute, very hot. Stem medium to tall, paler than cap. Grows with pines, especially in wet places. Common name: Rufous Milk-cap. Specific name, *rufus*, means 'red'.

## *Lactarius quietus*

Small to medium. Cap averages 3.5cm; dull reddish-brown with concentric darker and lighter zones and central depression. Gills paler than cap, slightly decurrent. Flesh pale buff, thickish in cap, often hollow in stem. Milk whitish, not changing; taste mild or slightly bitter. Stem similar in colour to cap but darkening towards the base. Smell oily, said to be of bugs. Grows with oak. Common name: Oak Milk-cap. Specific name, *quietus*, means 'calm'.

## *Lactarius tabidus*

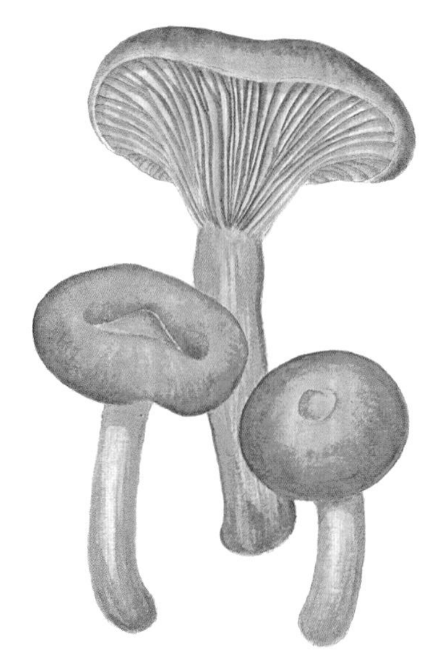

Small species. Cap averages 3.5cm; orange-buff, usually with central pimple. Gills yellower than cap, decurrent. Flesh whitish, rather thin. Milk white, turning yellow on a white cloth but not on the gills; slightly hot. Stem tall, similar colour to cap. Grows with deciduous trees, especially birch. Specific name, *tabidus*, means 'wasting away'. *L. subdulcis* is similar, but the milk does not change colour; usually found on beech.

## *Russula nigricans*

Large, thick and extremely brittle. Cap, which averages 10cm, gills and stem are all white to cream. Gills well spaced, adnate, intermediate gills present. Flesh hard and white; when cut, it turns first greyish-pink and then dark grey; taste hot. Eventually the whole fungus turns black but is slow to decompose, so entirely black specimens are most commonly found. Stem short and stout. Grows under many kinds of tree. Common name: Blackening Russula. Specific name, *nigricans*, means 'becoming black'. *R. densifolia* is similar but has gills much more closely packed. It, too, blackens all over.

### *Russula ochroleuca*

Medium-sized species. Cap averages 7cm; ochre-yellow. Gills pale cream, adnate. Flesh white, medium thickness. Stem of medium height, white, greying with age. It occurs with a wide variety of trees, both deciduous and coniferous, and is one of the most common Russulas throughout the season. The exact shade of yellow makes it recognisable in the field once it has been encountered a few times. Specific name, *ochroleuca*, means 'yellow and white'.

### *Russula claroflava*

Medium-sized species. Cap averages 7cm; bright yellow. Gills pale ochre, almost free of the stem. Flesh white, of medium thickness. Stem medium to tall, white. Grows under birches in wet, boggy, places. Can be differentiated from *R. ochroleuca* by its brighter yellow cap, easily recognised in the field, and by its situation. Specific name, *claroflava*, means 'bright yellow'.

## *Russula fellea*

Small to medium, sturdy Russula. Cap averages 5cm; honey-ochre. Gills paler than cap; adnate; taste very hot. Flesh almost white. Stem is a paler shade of the cap colour; stout and firm. The distinguishing feature of this fungus is its marked smell of geranium leaves. Grows under beech. Very common. Specific name, *fellea*, means 'full of gall'.

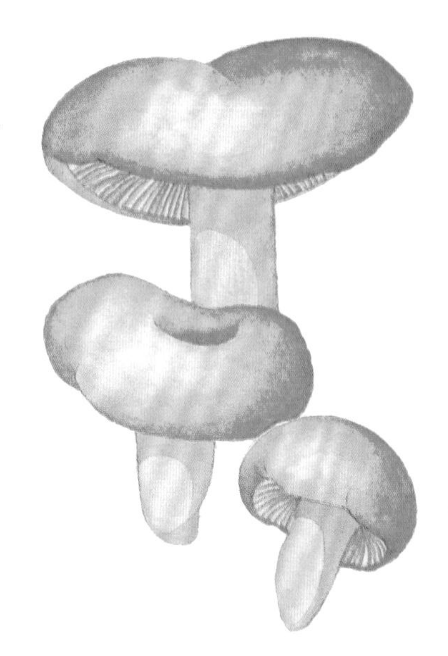

## *Russula cynoxantha*

Medium-sized species. Cap averages 7cm; may be almost any colour, but is typically dull lilac. Gills are white; adnate, crowded and sometimes forked; distinguished by the greasy, flexible feel when the finger is rubbed over them. Flesh white, of medium thickness. The stem, which is of average thickness is white and hard and occasionally has a flush of the cap colour on it. Common name: Charcoal-burner. Specific name, *cynoxantha*, means 'blue and yellow'.

## *Russula vesca*

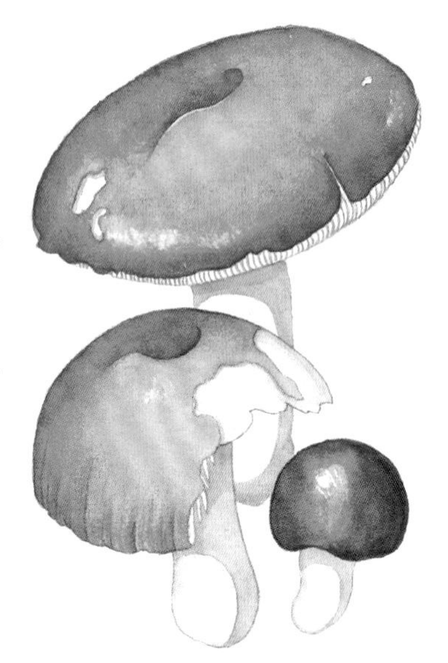

Small to medium. Cap thickish, averages 5cm across; of various colours, usually pale wine to buff. Cuticle tends to shrink from the margin, leaving the gill edges exposed. Gills white; adnate and closely packed; some show forking close to the stem. Flesh white. Taste mild, slightly nutty. Grows with deciduous trees. Common. Specific name, *vesca*, means 'I feed'.

## *Russula xerampelina*

Medium sized. Cap averages 7cm; ranging in colour from brown to purple, yellow or even green. Gills pale to medium ochre; adnate, fairly thick and deep, and connected by ridges where they join the cap. Flesh white, moderately thick. Stem stout, white, sometimes tinted pink. Distinguished by smelling of tinned crab, especially when old. Found mainly under beech and oak. Specific name derives from *xero*, meaning 'dry'.

## *Russula sardonia*

Medium sized. Cap averages 7cm; reddish to purple, sometimes brownish. Gills primrose to golden-yellow, turning rose-pink with a drop of household ammonia; adnate. Flesh thick. Stem stout, white usually overlaid with a strong purplish flush. Taste hot. Smell often of stewed apples. A common species where there is pine. Because of its variability it is worth testing all similar finds with ammonia. Specific name is derived from 'sardonyx', a reddish gemstone.

## *Russula atropurpurea*

Medium sized. Cap averages 7cm, dark, almost black, in the centre and reddish or purple round the sides. Gills pale cream, adnate, closely spaced. Flesh white. Stem white. Taste usually hot. Smell pleasant, of apples. Grows with oak or beech. Common name: Black-and-purple Russula. Specific name, *atropurpurea*, means 'black and purple'.

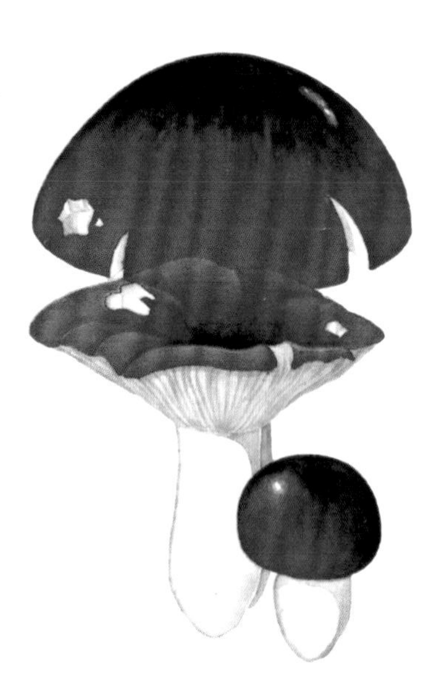

## *Russula foetens*

Large fungus distinguished by strong rancid smell. Cap averages 10cm; brownish-ochre, slimy, with a furrowed margin. Gills dirty cream; thick, spaced and adnate. Taste of the gills is very hot. Flesh white with cavities in stem. Stem short and stout; whitish. Specific name, *foetens*, means 'stinking'. *R. laurocerasi* is very similar in appearance; smells of marzipan or bitter almonds.

## *Russula mairei* **POISONOUS** 

Small. Cap averages 4cm; very bright red. Gills white, adnate. Flesh white. Stem medium to tall, tapering only slightly upwards. Taste hot. Grows under beeches. Specific name, *mairei*, is after Maire, a French mycologist. An apparently identical species, *R. emetica*, is found under pine. Both are poisonous in quantity. Common name for both species: The Sickener. *R. luteotacta* is a paler species with 'washed out', whitish areas on the cap; bruises yellow when handled or broken.

## *Panellus stipticus*

Small, thin-fleshed, shell-shaped bracket. Cap up to 3cm in width; pale cinnamon; thin with incurved margin; when torn apart the cuticle tears with it; breaks into fine scales (seen under magnification). Gills narrow. Stem short, lateral. Taste bitter and astringent. Grows on dead deciduous wood in tiers, often 20-30 or more in a group. Common, especially late in the season. Specific name, *stipticus*, means 'astringent'.

## *Crepidotus mollis*

Thin, watery, rubbery bracket. Cap usually small but may grow to 5cm across; pale cream; shell-shaped. Gills whitish becoming cinnamon. Flesh will break if slowly pulled apart but a thin elastic pellicle will remain: a good diagnostic test. Stem lateral. Found on dead branches or twigs of deciduous trees. Specific name, *mollis*, means 'soft'. The common *C. variabilis* is similar, but smaller, and has no elastic pellicle; usually grows on twigs or debris.

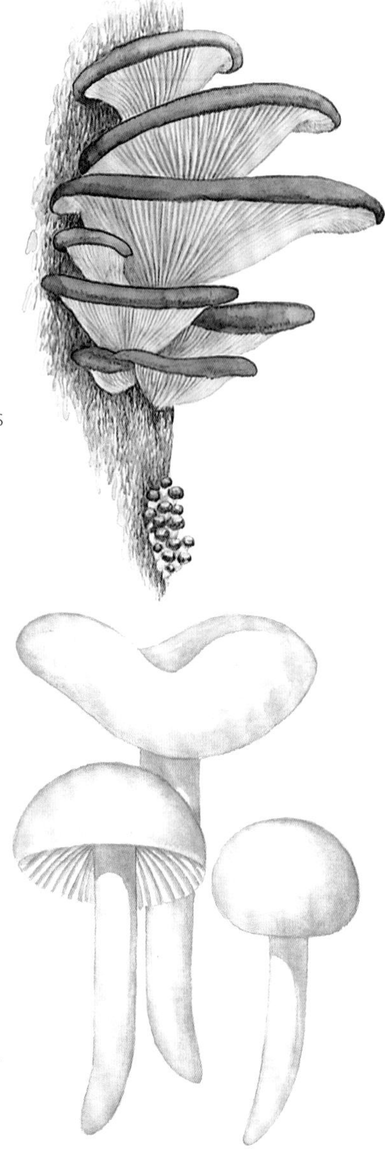

### *Pleurotus ostreatus*

Medium to large bracket of rubbery texture. Cap averages 7cm across; colour extremely variable, often starting very dark grey, becoming lighter, but may be pale buff from outset. Flesh white with white gills reaching down the very short lateral stem almost to the base, which is hairy. Spores very pale lilac. Edible and safe. Found on deciduous trees and logs, especially beech. Common name: Oyster Fungus. Specific name, *ostreatus*, means 'rough'. Similar species *P. cornucopiae* is usually creamy-white with the gills running down to the base, where they join and separate again. Spores white. Edible. Occurs as early as July.

### *Hygrophorus chrysodon*

Small, fairly stout agaric. Cap averages 4cm; white but yellowing with age; has an incurved margin, slightly squamulose. Gills white; adnate. Flesh white. Stem proportionately stout, basically white, but covered with fine yellow squamules also. Grows in deciduous woodland. Specific name, *chrysodon*, means 'gold tooth'. *H. eburneus* is similar but slightly more slender and lacks the yellow coloration; the cap is glutinous. *H. cossus* is larger, rarer and supposedly smells like the caterpillar of the Goat Moth (*Cossus cossus*), so-called because of its strong, goatlike smell.

## *Hygrophorus hypothejus*

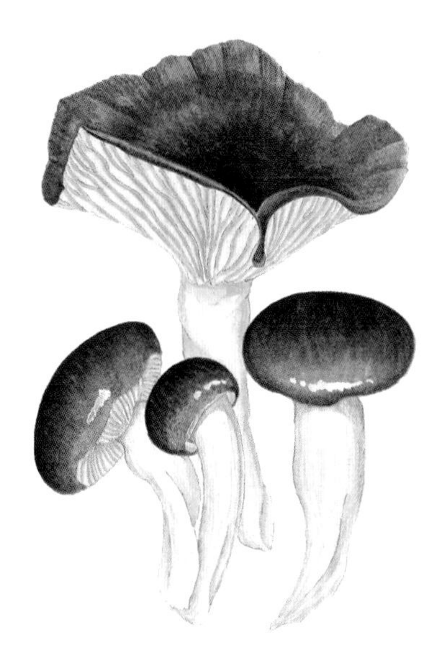

Small. Cap averages 3.5cm; dark greyish-brown but is covered with an olivaceous gluten, making it appear yellowish-orange. Gills decurrent, whitish, and slippery with orange gluten especially below the ring-zone. Stem relatively tall, white. Grows in pine woods. Occurs late in the season. Common name: Herald of Winter. Specific name, *hypothejus*, means 'under brimstone'.

## *Hygrocybe psittacina*

A small fungus. Cap, which averages 3cm, and stem are basically coloured cream to yellow but both are typically covered with a green slippery gluten which persists patchily all over. Gills adnate; white, though they often look greenish because of the gluten collected between them. Flesh whitish. Grows on grassland. Common name: Parrot Waxcap. Specific name is derived from *psittacos*, meaning 'parrot'.

## *Hygrocybe conica*

Smallish, waxy toadstool. Conical cap, which does not open, may start by being yellow, orange or even red, but very soon starts to turn black unevenly. Gills, which are well hidden at first, are whitish, sinuate. Flesh yellowish-white. Stem tall; coloured as the cap. Eventually the whole fungus becomes completely black, without, at first, losing its shape. Grows in grassland. Specific name, *conica*, means 'conical'. *H. nigrescens* is almost indistinguishable but has a white base to the stem. Occurs in grassy woodland clearings.

## *Hygrocybe punicea*

Medium-sized fungus. Cap bright blood-red at first, but the colour soon washes out with rain or frost leaving the surface with the appearance of being covered with a white bloom; rounded conical in shape. Gills are pale yellow, reddish at the base; adnate. Stem red with a white base; if split with a knife the flesh is white. Grows in grassland. Specific name, *punicea*, means 'blood-red'. The similar *H. splendidissima* also has a red tinge at base of gills, but the flesh of the stem is reddish.

## *Hygrocybe chlorophana*

A predominantly yellow fungus, despite its Latin name. Cap usually small, but some specimens may reach 5–7cm in diameter. Gills and flesh a consistent lemon-yellow, which is the diagnostic feature. Stem tall; smooth, slimy and often channelled. Grows in grassland. Specific name is derived from *chloros*, 'pale green', and *phana*, meaning 'I appear'. *H. konradii* has a lobed cap with a central umbo, and paler gills. Also a grassland species.

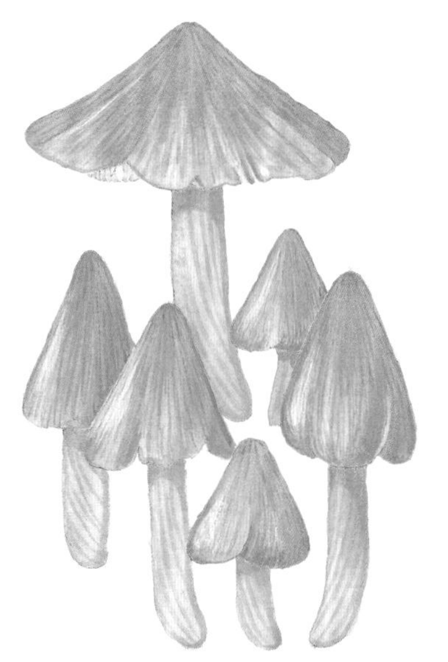

## *Camarophyllus niveus*

Commonest all-white, wax-gilled fungus. The cap is on average 2–3cm in diameter, rounded. Gills decurrent and well spaced. Stem is quite stout for the genus, though long in proportion to the size of the cap. The whole fungus tends to become more ivory-coloured with age. Grows in lawns and other grassland. Specific name, *niveus*, means 'snow-white'. *Hygrocybe russocoriacea* is similar but slightly smaller and more creamy. Smell strong and sweetly pleasant, said to be of Russian leather. Grows in short grass.

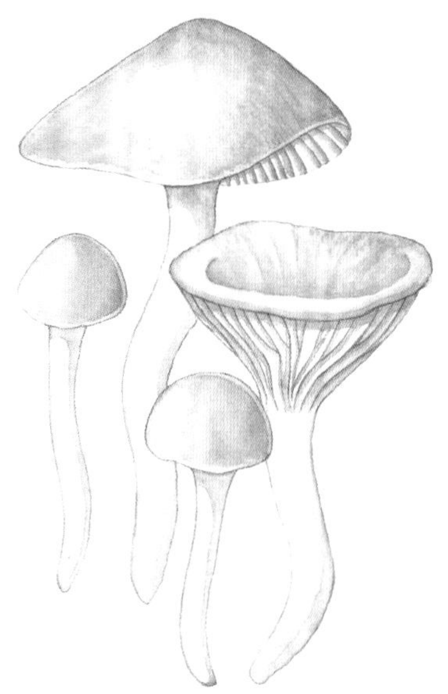

## *Camarophyllus pratensis*

Relatively large member of the genus. Cap may reach 8cm; bright orange-buff. Gills are pale buff; widely spaced, and decurrent. Flesh is quite thick at the centre, tapering towards the edge. The stem length is in proportion to the size of the cap. There is no other species with a bright orange-buff cap. Grows in unfertilised grassland, usually in small clusters. Specific name, *pratensis*, means 'growing in meadows'

## *Amanita phalloides* **POISONOUS** ☠

Medium size. Cap averages 7cm, whitish, darkened centrally by radiating fibres of green (or sometimes yellow). Gills white (differentiating this species from edible mushrooms); free. Flesh white. Stem tallish, white; it usually has a ring but often the veil that forms the ring remains hanging on the cap edge instead; the stem base is surrounded by a marked volva. Poisonous: eating one cap only can cause death. Occurs in woodland, usually under oak. Common name: Death Cap. Specific name, *phalloides*, means 'like a phallus'.

## *Amanita virosa* **POISONOUS** 

Fruitbody all-white. Cap averages 7cm. Gills white, free. Stem scaly or shaggy, with a volva and ring. Deadly poisonous. Grows in oakwoods. Less common in the south than *Amanita phalloides* but more common in the north and west. Common name: Destroying Angel. Specific name, *virosa*, means 'poisonous'.

## *Amanita muscaria* **POISONOUS** 

Most readily recognised of all toadstools. Cap averages 8cm in diameter; brilliant red, usually with white spots; however, the white spots may be lost and the colour washes out of the cap, leaving it orange. Gills white, free. Stem white, with a ring. Poisonous, though not as deadly as *A. phalloides* and *A. virosa*. Grows with birch. Common name: Fly Agaric, from its former use as a preparation for killing flies. Specific name, *muscaria*, means 'a fly'.

## *Amanita rubescens*

Sturdy, medium to large fungus. Cap averages 8–10cm in width when expanded; pale brown, covered with grey patches formed from the remains of the veil. Gills white, free. Flesh white. Stem white, turning pink, especially where damaged by slugs; has a ring which, on its outer surface, shows striation formed from contact with the gills; broadens towards base where it is ridged by remains of bottom part of volva. Edible, but can be confused with the poisonous *A. pantherina*, so should be avoided. Occurs in many types of woodland. Common name: The Blusher. Specific name, *rubescens*, means 'becoming red'.

## *Amanita pantherina* **POISONOUS** ☠

Medium to large. Cap averages 7cm; pale brown covered with white patches of volval remains (not grey, as *A. rubescens*). Gills, flesh and stem white. Stem does not turn pink when damaged. Poisonous and can be fatal, though not as toxic as *A. phalloides* or *A. virosa*. Grows in woodland. Common name: Panther Cap. Specific name, *pantherina*, means 'deceitful'. *A. excelsa* is similar in appearance, but the cap is grey-brown, the patches are grey and it has a series of ridges between the ring and the stem base.

## *Amanita citrina*

Medium sized. Cap averages 6cm in diameter; lemon-coloured with large white patches. Gills white, almost free. Flesh white. Stem fairly tall; has a ring and is swollen into a large bulb at the base. All parts of the fungus have a strong smell like that of raw potatoes, especially when cut. Grows in many types of woodland. Specific name, *citrina*, means 'lemon-coloured'. The variety *A. alba* is all-white, common.

## *Amanita fulva*

One of the harmless Amanitas. Cap averages 5–8cm in width; tawny, with marked striations radially placed all round the outer margin. Gills and flesh are white. Stem is relatively tall but with no ring, and has a large and conspicuous volva. Edible. Grows in woodland. Common name: Tawny Grizette. Specific name, *fulva*, means 'tawny'. *A. vaginata* is a similarly shaped, greyish-white fungus; less common. Common name: Grizette.

## *Macrolepiota procera*

Large, impressive mushroom. Cap can measure up to 25cm in diameter; buff with brown scales and parasol-shaped with a central umbo. Gills white, free. Flesh white. Stem brownish, covered with a darker layer, which breaks up into snake-like markings; carries a double ring which sits high up the stem but may be moved up and down as it has no attachment. Edible. Grows in grassland. Common name: Parasol Mushroom. Specific name, *procera*, means 'tall'. *M. rhacodes* is similar but has smooth stem. Edible. Found in grassland. Common name: Shaggy Parasol.

## *Lepiota cristata*

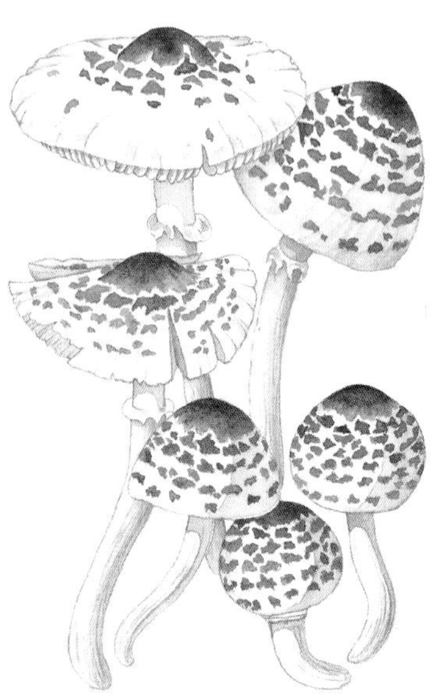

A small to medium fungus. Cap up to 7cm wide; white, covered with reddish-brown scales which are dense in the centre and absent towards the edge. Gills white; free and crowded. Flesh white. Stem up to 7cm long; slightly yellow or brown, carries a white or brownish ring. Inedible. Has an unpleasant smell. Grows in woodland and gardens. Specific name, *cristata*, means 'crested'.

## *Cystoderma amianthinum*

Small fungus, at one time in the genus *Lepiota*. Cap averages 3cm; orange-brown and scaly. Gills white, adnate. Flesh white. The stem is only slightly longer than the width of the cap; it is white above the immovable ring but both the underside of the ring and the whole of the stem below it are covered with scales similar in colour to the cap. Occurs in moss or short grass. Specific name, *amianthinum*, means 'unspotted'.

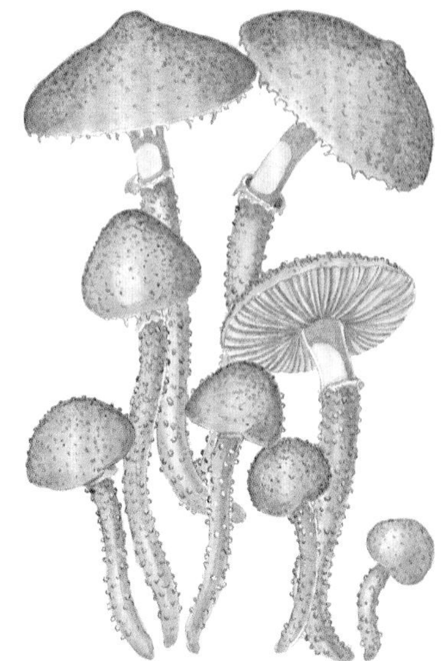

## *Armillaria mellea*

Medium-sized fungus. Cap averages 9cm in diameter; often honey-coloured though this is variable; slightly scaly. Gills white, slightly decurrent. Flesh white. Stem, similarly coloured to the cap, bears a white ring. Grows in tufts; lower caps are often coloured with white spores shed by the higher ones. Grows on trees, which it often destroys. Common name: Honey Fungus. Specific name is derived from *mel*, meaning 'honey'. Similar species include *A. bulbosa*, with a paler, thicker stem; *A. ostoyae* with black scales on cap and ring; and *A. tabescens* which lacks ring. All grow in woodland.

## *Oudemansiella mucida*

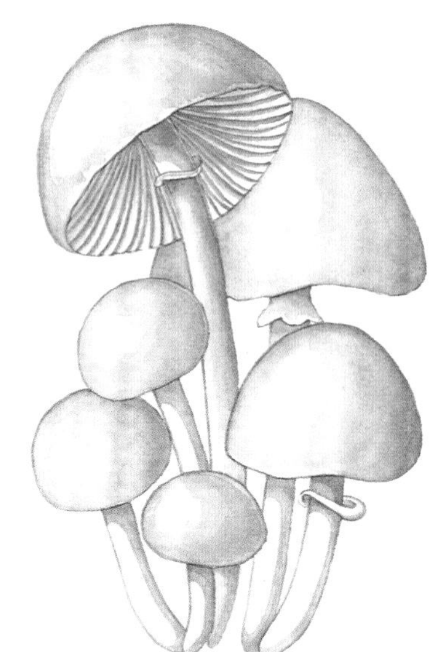

Medium-sized, all-white fungus. Cap very variable in size; slimy; becomes translucent. Gills free. Stem length varies considerably as it rises to clear the cap from the tree on which it grows; slightly scaly and carries a membranous ring. It may occur both on fallen branches and high up the tree. Common name: Porcelain Fungus. Specific name, *mucida*, means 'covered in slime'.

## *Oudemansiella radicata*

Medium-sized fungus. Cap averages 7cm in diameter; brown, radially furrowed. Gills white, free. Flesh white. Stem tall, twice cap width; white, shading into brown at the base. If carefully dug up, a long extension of the stern, which attaches it to buried wood, can be seen. Usually grows on beech. Common name: Rooting Shank. Specific name, *radicata*, means 'rooting'.

## *Collybia maculata*

Medium-sized fungus. Cap averages 10cm in diameter; white or cream. Gills free. Stem may be elongated, according to depth of medium in which it grows; colour as cap. All parts, especially cap and gills, develop brownish spots with age. Grows in leaf litter of either deciduous or coniferous woods, or in bracken litter. Common name: Spotted Tough-shank. Specific name, *maculata*, means 'spotted'.

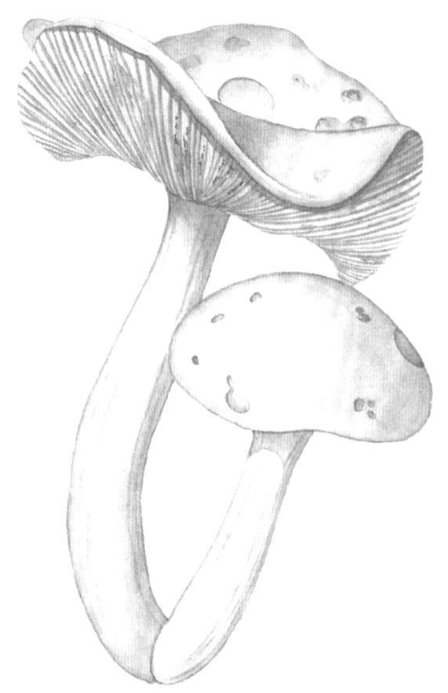

## *Collybia fusipes*

Small to medium-sized fungus. Cap averages 6cm; brown and smooth. Gills fawn-coloured and free. Flesh fawn-coloured. Stem characteristic, red-brown, longitudinally grooved, widening and darkening towards the centre and tapering towards the base, which forms quite a long 'root'. Grows on oak and is usually found close to the base of the tree. Common name: Spindle-shank. Specific name, *fusipes*, means 'spindle-foot'.

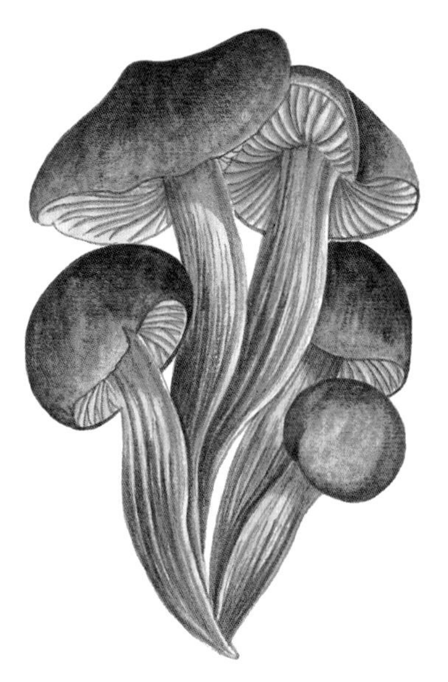

## *Collybia dryophila*

Small, very variable fungus, responsible for much of the leaf decomposition in deciduous woodland. The cap is thin, average width 4cm, variable in colour but usually pale. Gills white, free. Flesh white, thin. Stem thin, long and hollow, widening and darkening towards the base. Grows in all types of woodland throughout the season. Very common. Specific name, *dryophila*, means 'oak-loving', though it is not confined to oak.

## *Collybia peronata*

Medium sized. Cap averages 6cm; pale grey to brown. Gills become yellow with age; free. Flesh creamy. The stem is fairly long, depending on the depth of the leaf litter in which it grows; pale, yellowing with age; the base is covered with hair. Grows characteristically in deciduous woods, but may occur with conifers. Common name: Wood Woolly-foot. Specific name, *peronata*, means 'wearing woolly boots'.

## *Marasmius oreades*

Small to medium-sized fungus. Cap pale, averages 5cm. Gills white, adnate, widely spaced. Flesh white. Stem medium length, of similar colour to cap. Like other species of this genus, it has the ability to dry out and later rehydrate without deteriorating. Edible but tough, not recommended because it can be confused with similar poisonous species. Grows in rings on cultivated lawns. Common name: Fairy Ring Champignon. Specific name, *oreades*, means 'of mountains'.

## *Laccaria laccata*

Small, highly variable fungus. Cap tan coloured, but extreme variability of colour and shape cause it to be misidentified frequently when seen from above. Gills characteristically flesh-coloured, well spaced and adnate. Stem tall in relation to the size of the cap; fibrous and twisted. Grows in almost any situation from deciduous and coniferous woods to cultivated lawns. Common name: The Deceiver. Specific name, *laccata*, means 'resinous exudation'. Similar species are *L. amethystea*, which is identical in shape but an overall dull amethyst colour, and *L. bicolor* which has amethyst gills and stem base.

## *Calocybe gambosum*

Formerly known as *Tricholoma georgii*; medium sized, stout, resembling a cultivated mushroom. Cap white. Gills white; adnate and crowded. Flesh thick. Smells strongly of new meal. Edible. Grows in rings on lawns and meadows. Common name: St George's Mushroom; it is supposed to appear on St George's Day (23 April) but is usually a month later. Specific name, *gambosum*, means 'resembling a hoof'.

## *Tricholoma fulvum*

Medium sized. Cap averages 7cm in width; tawny, usually with slightly darker streaks radiating round the edge. Gills yellow, spotting brown with age; adnate. Flesh yellow. Stem tallish, similarly coloured to cap but paler at the top. Grows in woodland, usually associated with birch. Specific name, *fulvum*, means 'tawny'. Similar species include *T. sulphureum*, slightly smaller, yellow, smelling of coal tar or creosote, and *T. saponaceum*, larger, greyish, with scaly or spotted stem, smelling of soap.

## *Tricholomopsis rutilans*

Medium to large fungus. Cap averages 7cm; yellow, but the colour is almost completely obscured by a layer of red-brown scales. Gills bright egg-yellow; adnate. Flesh cream. Stem paler than cap. Occurs with conifers, especially pine. Common name: Plums and Custard. Specific name, *rutilans*, means 'becoming reddish'.

## *Tricholomopsis platyphylla*

Medium-sized fungus. Cap averages 8cm in width; grey-brown with radiating fibrils. Gills white; broad and adnate. Flesh white. Stem variable in height; white but covered with brownish fibrils. Grows in the leaf litter of deciduous woods; if dug up carefully will be seen to have long white mycelial cords attached to the base of the stem. Specific name, *platyphylla*, means 'broad leaf'.

## *Lepista nuda*

Medium sized and sturdy. Cap averages 8cm in width. The whole fungus – cap, adnate gills, and stem when young and fresh – is a bluish-lilac colour but the surface of the cap rapidly fades and becomes almost buff. The stem tends to broaden towards the base. Edible. Occurs in almost any position but is more often associated with deciduous trees. A late species, usually occurring in November or December. Common name: Wood Blewit. Specific name, *nuda*, means 'naked'.

## *Lepista saeva*

Medium-sized species, similar to *L nuda*. Cap averages 8cm across; pale tan coloured. Gills coloured as cap; adnate. Stem lilac, unchanging; usually swollen. Smell perfumed, strong and pleasant – perhaps the most distinguishing feature. Edible. Grows in grassland. Fruits earlier than *L. nuda* and is much less common. Common name: Field Blewit. Specific name, *saeva*, means 'wild'.

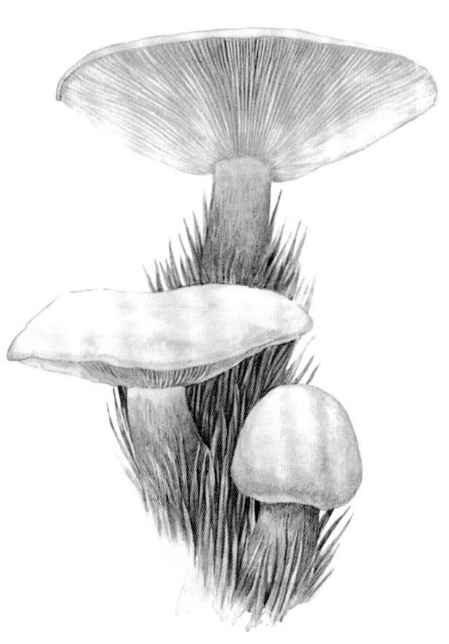

## *Melanoleuca melaleuca*

Medium sized. Cap averages 7cm in diameter; dark brown, becoming paler with age; usually shows a central umbo. Gills white; sinuate. Flesh white. Stem fairly tall, white with greyish fibrils, becoming darker with age, so retaining contrast with the cap throughout the life of the fungus. Usually grows in grassy woodland. Specific name, *melaleuca*, means 'black-and-white'. There are several very similar species.

## *Clitocybe nebularis*

Medium to large. Cap variable but averages 10cm in diameter; grey, thick. Gills creamy; decurrent. Flesh whitish. Stem varies in length; sturdy, widening towards the base. Has a strong, sweet smell. The young caps are edible but they can cause gastric disorders. Grows in clusters, mainly in deciduous woodland. Appears late in year. Common name: Clouded Agaric. Specific name, *nebularis*, means 'clouded'.

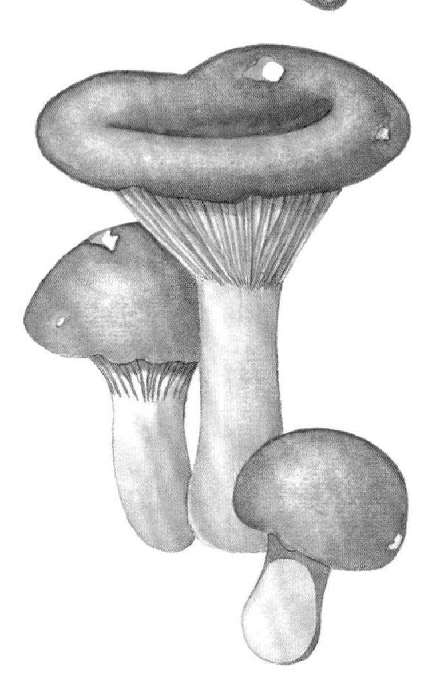

## *Clitocybe clavipes*

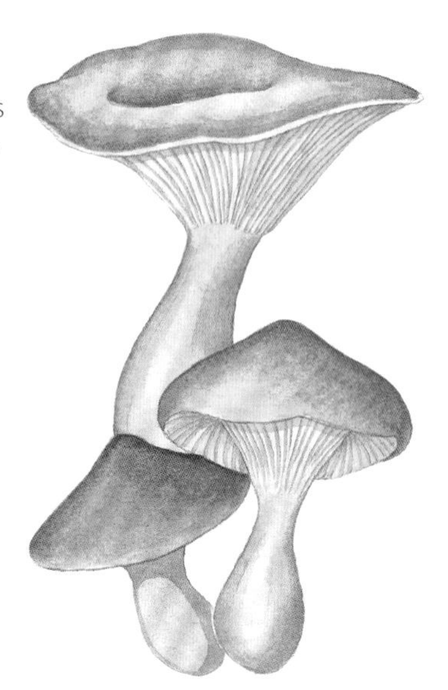

Small to medium species. Cap averages 5cm in diameter; grey-brown in colour, rounded or flat; it retains an incurved edge. Gills creamy-white, decurrent. Flesh white. Stem is swollen towards the base. Occurs mainly with beech and birch. Common name: Club-foot. Specific name, *clavipes*, means 'club-foot'.

## *Clitocybe infundibuliformis*

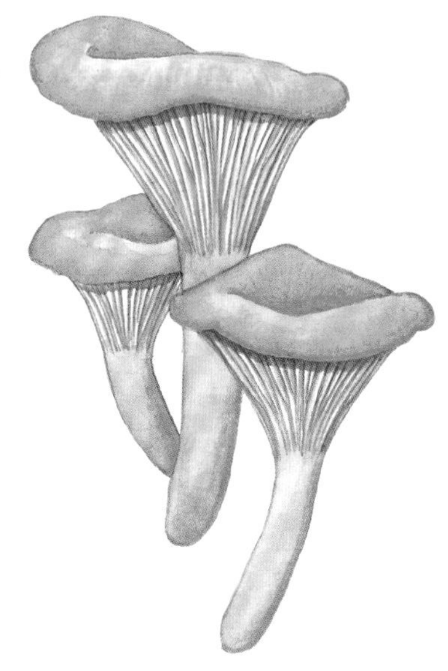

Small to medium fungus. Cap averages 6cm and is a pale wash-leather colour; funnel-shaped, smooth and silky. Gills creamy-white, running down the parallel-sided stem, which is short and slim. Grows in deciduous woodland. Common name: Funnel Cap. Specific name, *infundibuliformis*, means 'funnel-shaped'. *C. flaccida* is similar in every way except that it is a tawny colour.

## *Clitocybe dealbata* **POISONOUS** ☠

Medium-sized, rather mushroom-like fungus. Cap averages 6cm in diameter; white. Adnate gills, flesh and stem are also all-white. It has a whitewashed or chalky appearance. Grows in groups or rings and is rather slimmer than a mushroom. Extremely toxic and has caused death by being mistaken for a mushroom and eaten. Occurs in grassland. Not common. Specific name, *dealbata*, means 'whitewashed'.

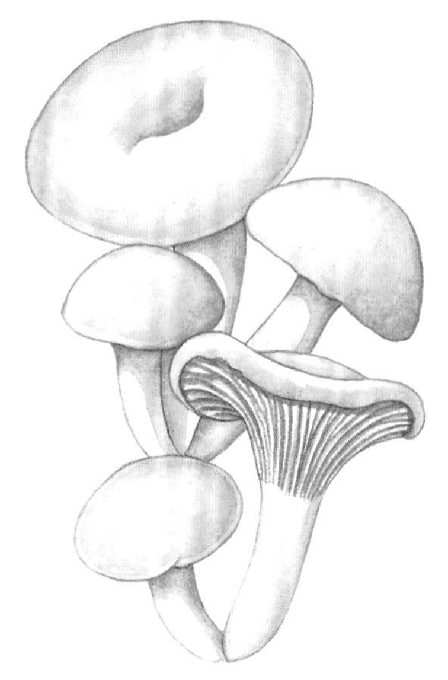

## *Hygrophoropsis aurantiaca*

Although now in a different genus, this small species is very similar in shape to *Clitocybe*. Cap averages 5cm across; dark orange. Gills orange, crowded and very decurrent. Flesh and stem also orange. It is not poisonous but has little flesh and no flavour. It grows under conifers, especially pines. Common name: False Chanterelle, due to its superficial resemblance to the Chanterelle. Specific name, *aurantiaca*, means 'orange coloured'.

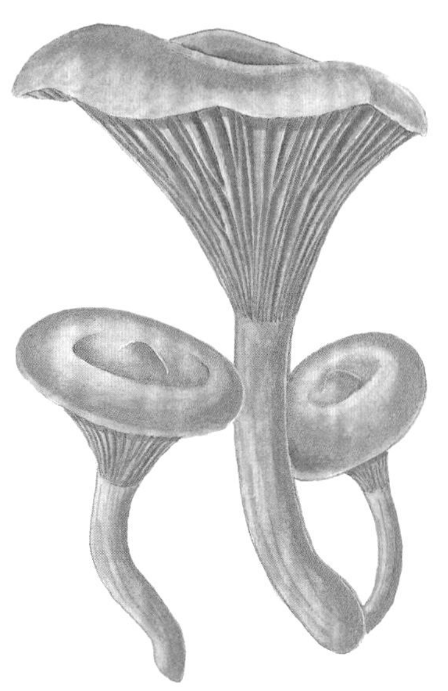

## *Strobilurus esculenta*

Small species. Cap usually averages about 2.5cm in diameter; pale, greyish-white. Gills whitish, almost free. Stem rather thin; similarly coloured to cap. Grows on spruce cones, usually those that are partly buried in the ground. Specific name, *esculenta*, means 'edible'. *S. tenacella* is a darker species that grows from buried pine cones. Careful excavation may be needed to find the root connection.

## *Mycena galericulata*

Generally a small species. Cap may sometimes reach 5 or 6cm across; grey, drying brownish. Gills adnate, white, though developing a flesh-pink tint with age. Flesh white. Stem long, similar colour to cap. Grows in tufts on deciduous wood. Specific name is derived from *galericulum*, meaning 'a cap'. *M. inclinata* is similar but has a browner cap and a reddish stem. Found on oak logs and stumps.

## *Mycena pura*

A small species, usually solitary or growing in small groups. Cap averages 3cm in diameter; may be either pinky-brown or grey-brown. Gills white; adnate. Flesh white. Stem white, rather stouter and shorter than others in the genus. Can be recognised by smelling more or less strongly of radish. Grows in deciduous leaf litter. Specific name, *pura*, means 'clean'.

## *Mycena galopus*

A small species. Cap averages 1.5cm; pale umber-grey, darker in the centre. Adnate gills and flesh white. Stem long and thin; grey, with white cottony base; oozes white milk when cut or broken. Grows in tufts on deciduous leaf litter. Specific name, *galopus*, means 'milk-foot'. *M. leucogala* also yields white milk but is very dark or nearly black in colour.

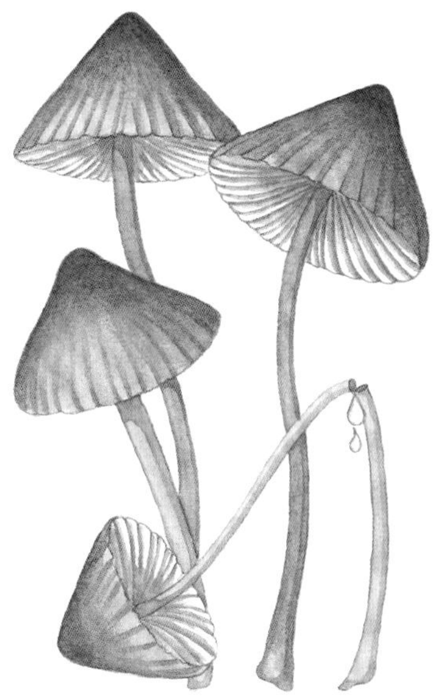

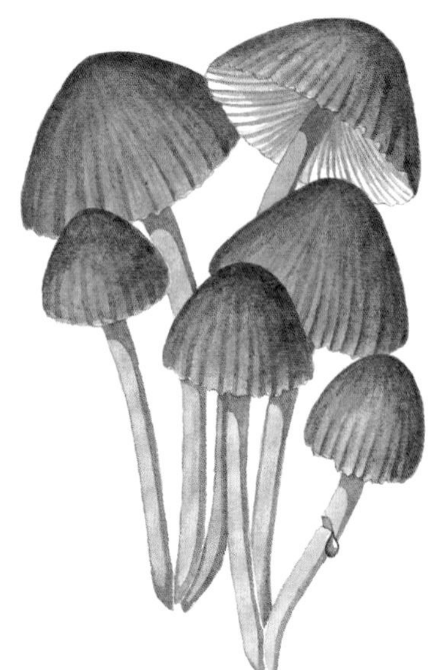

## *Mycena haematopus*

A small, tufted species. Cap averages 1.5cm; pinkish-brown, drying much paler but often showing streaks or blotches of darker colour; conical. Adnate gills and flesh white when undamaged. The stem is of similar colour to the cap and when cut oozes drops of blood-like sap. Grows on dead deciduous wood and stumps. Specific name, *haematopus*, means 'bloody foot'. *M. sanguinolenta* is a smaller version, recognisable by the red edges to the gills. Grows on the ground or on twigs.

## *Mycena epipterygia*

Relatively tall species. Cap averages 2cm; pale grey or yellowish. Gills dirty white; adnate. Flesh coloured as gills. Stem tall in relation to size of cap; pale grey, covered with a layer of yellowish gluten; if picked up it will adhere to the fingers. Grows in leaf litter or bracken. Specific name, *epipterygia*, means 'upon a small wing'.

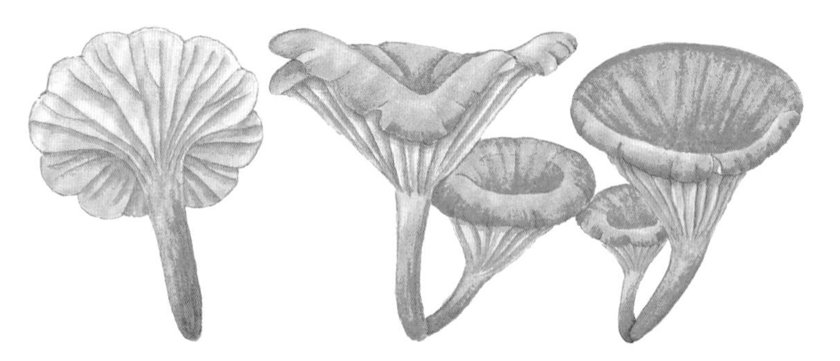

## *Omphalina ericetorum*

This is typical of the *Omphalina* genus. Cap averages 2cm; off-white, becoming yellowish or pale brown with age; characteristically funnel-shaped. Gills white; spaced and markedly decurrent. The stem is short, thin and similarly coloured to the cap. It is the fruitbody of the fungus component of a small green lichen which occurs on damp, peaty soil. Specific name, *ericetorum*, means 'of heaths'.

## *Pluteus cervinus*

Medium sized. Cap averages 10cm in width; varying shades of brown. Gills white, though the pink colour of the spores shows in the spaces between them; free. Flesh white. Stem tallish, white with brownish fibrils. It grows on rotting wood, especially beech. Specific name, *cervinus*, means 'colour of deer'. Similar in appearance to *Tricholomopsis platyphylla*, which grows in leaf litter and can be distinguished by its long white root-fibrils.

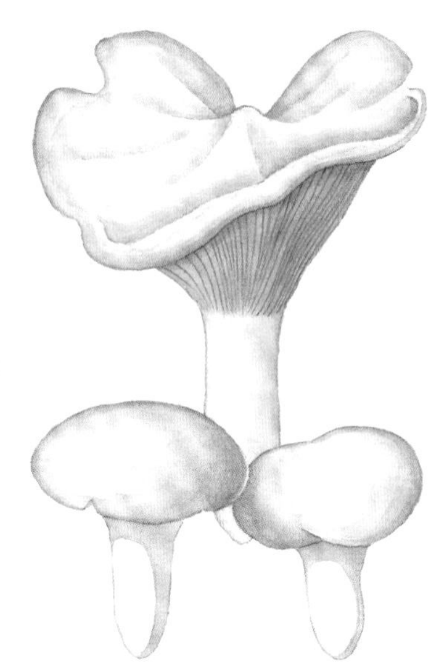

## *Clitopilus prunulus*

Small to medium-sized fungus. Cap averages 5cm across but widely variable; creamy-white. Gills white but, as the spores develop, show a pinkish tint; closely spaced and decurrent. Flesh white. Stem short, white. Distinguished by a strong smell of meal. Grows in woodland. Common name: The Miller. Specific name, *prunulus*, means 'small plum'.

## *Entoloma sinuatum* **POISONOUS** ☠

Sturdy, medium-sized species. Cap averages 7cm in diameter; creamy-white; may become waved at edge. Gills adnate; they start white and become yellowish-pink as the coloured spores develop. Flesh white, fairly thick. Stem white. Poisonous. It grows in deciduous woodland. Specific name, *sinuatum*, means 'waved'.

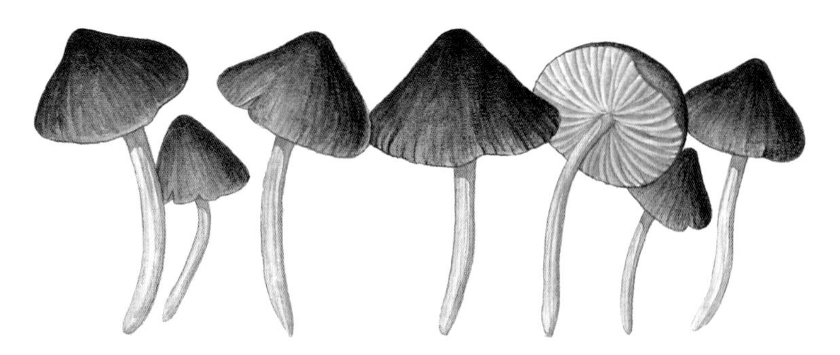

## *Nolanea staurospora*

Small to medium-sized species. Cap brown, broadly conical. Adnate gills grey at first but they become pink as the spores develop. Flesh pale grey. Stem tall and slim; grey. It is common in all types of grassland. Specific name, *staurospora*, means 'having cross-shaped spores'. A shorter, similarly coloured species with a silky cap is *N. sericea*, which grows in lawns.

## *Pholiota squarrosa*

Tufted, medium-sized species. Cap averages 8cm in diameter; deep yellow and covered in dark brown scales. Gills yellow; adnate; shed rusty brown spores. The stem is long in relation to the size of the cap; colour and scales as cap. Grows on deciduous trees, especially beech. Specific name, *squarrosa*, means 'scaly'. Related species *P. aurivella* has a slimy cap and is much less scaly. Also grows on deciduous trees.

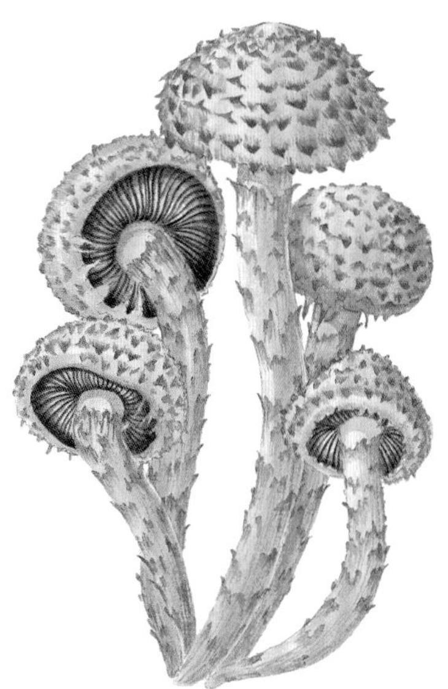

## *Gymnopilus junonius*

Large, tufted species. Cap averages 12cm in diameter; deep golden yellow. Gills are adnate; yellow, coloured by rusty spores, which fall on the stem. Flesh yellow. Stem coloured as cap; stout and swollen towards the base. It grows on deciduous wood, often buried so the fungus appears to be growing on the ground. Specific name, *junonius*, means 'belonging to Juno'.

## *Galerina mutabilis*

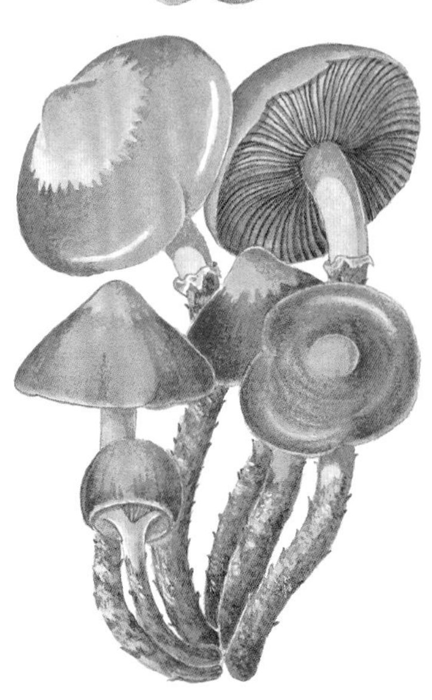

Now reclassified as *Kuehneromyces mutabilis*, but the older and better-known name has been retained here. Small to medium-sized species. Cap averages 5cm in diameter; brown, recognised by the fact that it changes colour as it dries, becoming pale in the centre. Gills pale cinnamon-brown; adnate. Flesh whitish-cinnamon. The stem is smooth and pale above the ring, dark and scaly below. Grows on deciduous wood. Specific name, *mutabilis*, means 'changing'.

### *Cortinarius pseudosalor*

One of a great many species of *Cortinarius*, few of which can be recognised in the field. Cap averages 8cm in diameter when expanded; rich brown and slimy, paler and striate on outer third. Gills are rusty coloured; adnate. Flesh pale brown. The stem is slimy, slightly swollen in the middle; pale blue, shows a darker ring zone where the cap edge separated from it. Grows mainly in deciduous woods, especially on beech. Specific name is a compound of *pseudo* meaning 'false' and *salor* meaning 'from the high sea'.

### *Cortinarius speciosissimus* **POISONOUS** ☠

Medium sized. Cap averages 7cm in diameter; tawny to date-brown, darker in the centre; dry and fibrous. Adnate gills and flesh similarly coloured. Stem tall; colour similar to cap; dry and fibrous. Spores rusty-coloured. Extremely poisonous and can destroy the kidneys. Grows under pines. Northern distribution. Specific name, *speciosissimus*, means 'most handsome'. Similar species are found growing under deciduous trees and in more southerly areas. All are poisonous.

## *Bolbitius vitellinus*

Small to medium species. Cap averages 4cm in diameter, bell-shaped; creamy-white developing a yellow centre; it is at first viscid, then furrowed and splitting at the margin. Gills adnate; at first yellowish, then becoming rusty. Stem tall and slim, tends to thicken at the base; creamy and at first covered with a pale meal. Grows on dung or in grass enriched by dung. Specific name, *vitellinus*, means 'like egg-yolk'.

## *Paxillus involutus*

Medium sized species, related to the *Boleti*. Cap averages 9cm in diameter; downy brown; it maintains an inrolled margin for a long time, then becomes funnel-shaped and reveals the decurrent, ochre gills. These are readily pushed off the stem with a fingernail. The flesh and stem are similar in colour to the cap. Occurs in many kinds of woodland, but is usually associated with birch. Specific name, *involutus*, means 'inrolled'.

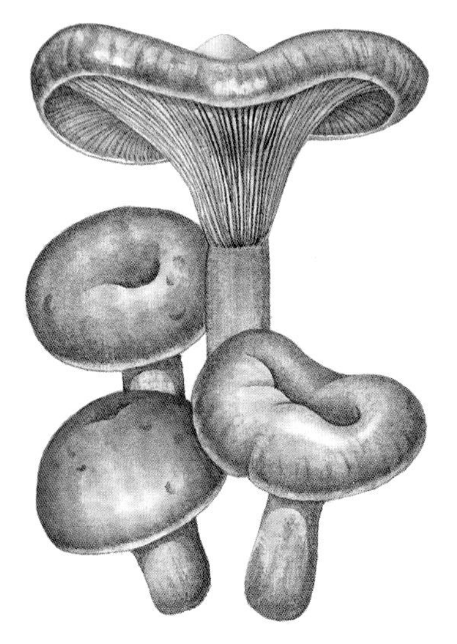

## *Inocybe geophylla* POISONOUS ☠

Small fungus. Cap averages 2cm in diameter; white; at first conical and then as it opens out it becomes umbonate and silky. Gills almost free, at first whitish and then clay-coloured; the spores are rusty. Flesh white. The stem is even, fairly long relative to the cap and often bent. Poisonous. It is found in all types of woodland. Specific name, *geophylla*, means 'earthleaf'. There is a lilac-coloured form, var. *lilacina*.

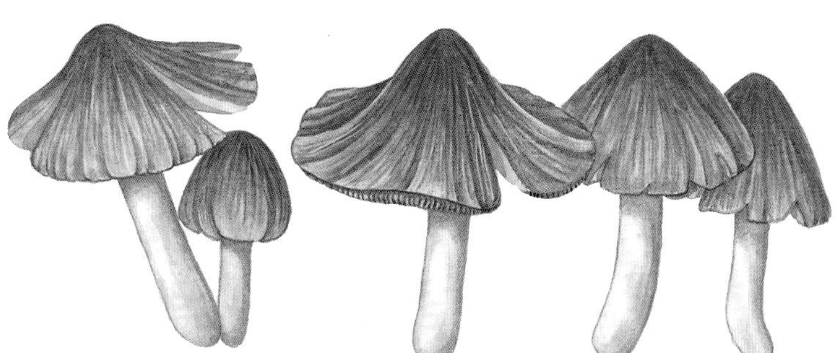

## *Inocybe fastigiata* POISONOUS ☠

Small to medium fungus. Cap averages 4cm, reminiscent of a bell-tent in shape; pale yellowish, characteristically splitting radially and revealing the pale flesh underneath. Gills are dirty yellow with a white edge; crowded and adnate. The stem is even, fairly tall and slim and of similar colour to the cap. Poisonous. Found under deciduous trees, especially beech. Specific name, *fastigiata*, means 'gabled'.

## *Inocybe patouillardii* POISONOUS ☠

Medium-sized fungus. Cap averages 8cm in diameter; ivory-coloured with red radial fibres; silky and bluntly pointed. Gills are at first pink and then grey-brown, bruising red; adnate. Stem of medium height, stout; ivory-coloured as cap, with red fibrils running down. All parts bruise red. Deadly poisonous. It is characteristic of beechwoods on chalk. Summer species. Uncommon. Specific name is after Patouillard, a French mycologist.

## *Hebeloma crustuliniforme* POISONOUS ☠

Small to medium species. Cap averages 4cm in diameter; pale tan darkening in the centre almost to a dark brick colour; sticky and long remaining inrolled at the edges. Gills are first a clay colour and eventually date-brown; adnate; in wet weather they develop drops of water on the edges, which catch the spores and dry to leave brown spots. Flesh white and thick. Stem is pale whitish-fawn. Toxic. Found in deciduous woodland. Common name: Poison Pie. Specific name, *crustuliniforme*, means 'cake-shaped'.

## *Hebeloma sinapizans* POISONOUS ☠

Medium sized species. Cap averages 8cm in diameter, yellowish-clay coloured, darkening in time to pale tan; sticky. Gills are cinnamon coloured and the spores rusty; crowded and adnate. Flesh white. The stem is pale, sturdy and has a basal swelling. Smells quite strongly of radish. Poisonous. Grows in deciduous woodland. Specific name, *sinapizans*, means 'mustard'.

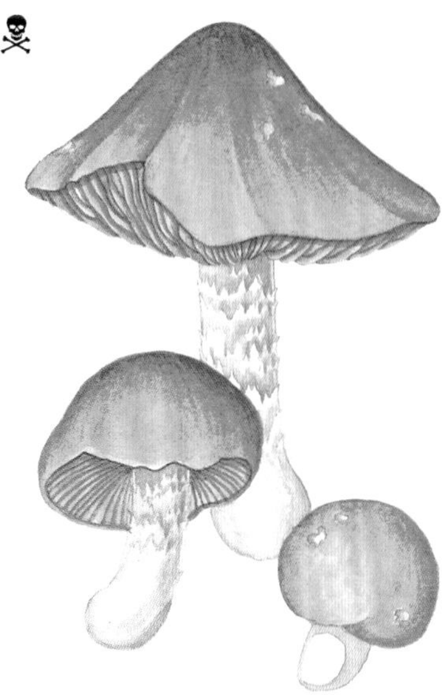

## *Agaricus campestris*

The well-known Field Mushroom; medium size and sturdy. Cap averages 6cm; creamy-white, with variable brownish scales on the surface. Gills start pink before becoming black; adnate. Flesh white. The stem is white with a ring. Edible and of excellent flavour. It occurs in fields where animals are grazed. Specific name, *campestris*, means 'of the plains'.

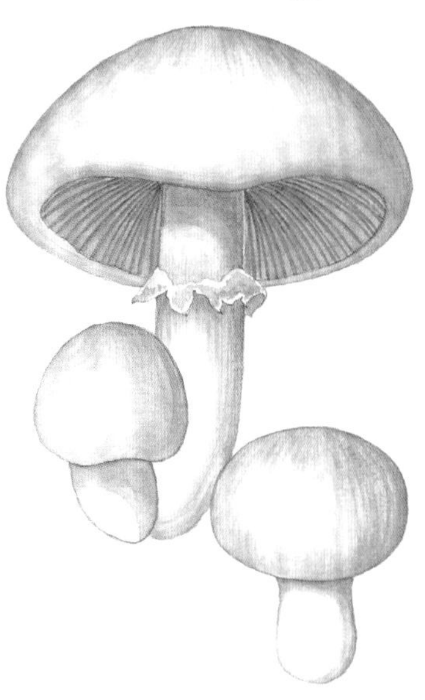

## *Agaricus arvensis*

Large fungus. Cap averages 15cm in diameter when fully open; ivory-white stained with yellow, mealy when young, then silky. Gills commence white, becoming greyish-pink and then black; adnate and almost free. Stem stout, carrying a thick, two-layered ring. Edible and even more tasty than the Field Mushroom. It occurs in cultivated fields, often in large 'fairy rings'. Common name: Horse Mushroom. Specific name, *arvensis*, means 'of cultivated fields'.

## *Agaricus xanthoderma* **POISONOUS** ☠

Often mistaken for Field Mushroom. Cap white, similar in shape to the cultivated mushroom. Gills whitish to pale pink, becoming grey then black; adnate to free. Flesh at the base of the stem discolours yellow when it is cut. Stem ringed. Smell unpleasant, inky. Inedible, can cause quite severe digestive disturbances. Occurs in cultivated fields. Common name: Yellow Staining Mushroom. Specific name is derived from *xanthoderma*, meaning 'yellow skin'.

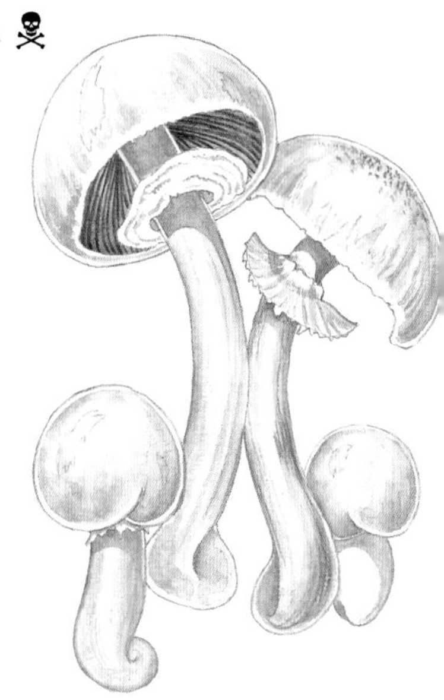

## *Agaricus silvicola*

Medium-sized species. Cap averages 9cm in diameter; yellowish; smooth and shining. Gills are never pink but become greyish; adnate. Flesh thin. Stem has a slightly bulbous base and carries a large ring. Smell pleasant, of anise. Edible and similar in taste to Field Mushroom. Found in woodland. Grey-capped species, also found in woodland, should not be eaten. Specific name, *silvicola*, means living in woods'.

## *Stropharia aeruginosa* **POISONOUS** ☠

Small to medium species. Cap averages 6cm in diameter; characteristically coloured verdigris green; slimy. Gills are initially white but become grey, then purplish-black; adnate to sinuate. Flesh bluish-white. Stem white, squamulose, with a small but definite ring, which catches the spores and becomes black. Poisonous. Grows in woodland. Specific name, *aeruginosa*, means 'stained by copper salts'.

## *Stropharia semiglobata*

Small, extremely common species. Cap averages 3cm; pale yellow, hemispherical. Gills black, wide, attached to the stem, forming a diameter to the hemisphere. The stem is of similar colour to the cap and carries a slight black ring. Found on dung, especially of horse. Common name: Dung Roundhead. Specific name, *semiglobata*, means 'half a globe'. The only similar fungus is *Panaeolus semiovatus*.

## *Hypholoma fasciculare*

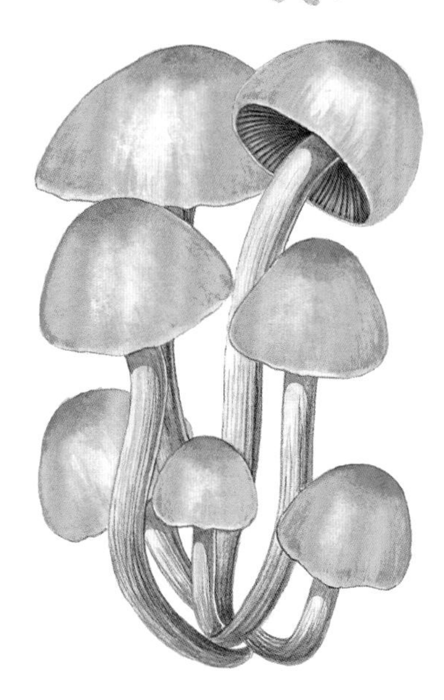

Small species. Cap averages 5cm in width; sulphur-yellow when young, with a slightly darker disc. Gills initially yellow, developing a green tint and then blackening; adnate. Flesh yellowish. Stem coloured as cap; usually curved. Grows in tufts, the lower caps becoming discoloured from the black spores of those above. Found on dead wood of any type. Common name: Sulphur Tuft. Specific name, *fasciculare* means 'in little bundles'.

## *Lacrymaria velutina*

Medium-sized species. Cap averages 8cm; pale ochre-brown, convex, covered in woolly fibrils, which overhang the edge as a fringe. Gills are dark purplish-brown with a white edge; adnate; characteristically covered with droplets. The stem is white at the top becoming brown below the ring-zone; scaly towards the base. Found on grassy roadsides. Common name: Weeping Widow. Specific name, *velutina*, means 'velvety'.

## *Psilocybe semilanceata*

Small species. Cap averages 1–1.5cm; characteristic bonnet shape, with a pointed tip and an incurved margin; it is covered in a glutinous pellicle. Gills black, adnate. The stem is long and thin, of pale buff colour. It occurs in short grass everywhere. Common name: Liberty Cap; though more recently its hallucinogenic properties have given it the name of Magic Mushroom. Specific name, *semilanceata*, means 'half spear-shaped'.

## *Coprinus comatus*

Medium-sized species. Cap averages 6cm; white, becoming brown on top; oval; hairy-scaly, at an intermediate stage the hairs turning outwards. Gills free, becoming black with spores and liquefying so that the spores are spread by rain. Flesh white when young. Stem tall, white when young. Grows on buried wood. Common names: Lawyer's Wig; Shaggy Ink Cap. Specific name, *comatus*, means 'hairy'.

## *Coprinus atramentarius*

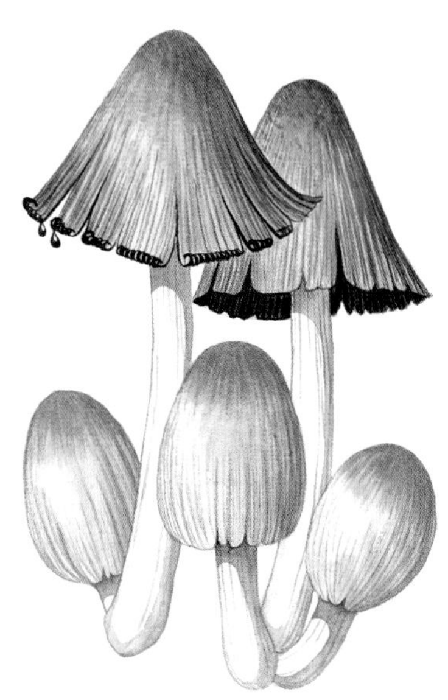

Medium-sized, normally tufted species. Cap averages 7cm in diameter; greyish-white, turning black; egg-shaped, fails to open; furrowed. Gills free; white becoming black and liquefying into 'ink', which was once used for writing. Stem tall, fairly slim. Edible, but contains a chemical related to 'antabuse' which will cause sickness if eaten in conjunction with alcohol. Grows in a wide variety of habitats, and is capable of pushing its way through paths and paving. Specific name, *atramentarius*, means 'inky'.

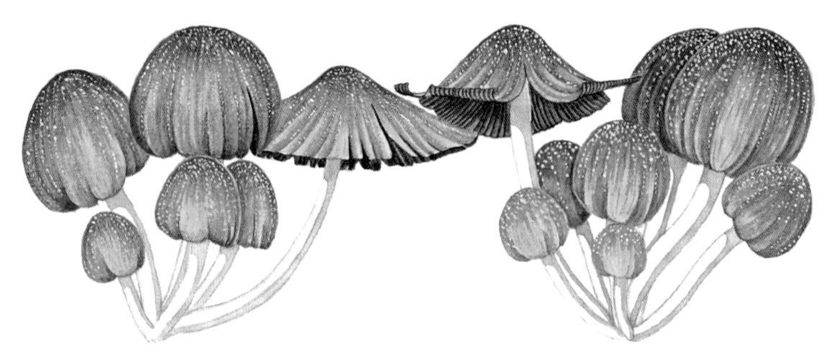

## *Coprinus micaceus*

Small species. Cap averages 4cm; pale ochre-yellow, darker towards the centre, ornamented on the top with glistening flecks of veil; soon darkens with age and loses the micaceous granules; oval when young, becoming bell-shaped. Gills white then blackening; free. Flesh whitish-grey. Stem white and smooth. Grows on wood, in dense clusters, or sometimes on the ground attached to buried wood. Specific name, *micaceus*, means 'like mica'.

## *Panaeolus sphinctrinus*

Small species, which undergoes an extreme change of colour as it dries. Cap averages 3cm; blackish-brown when young, changing to a pale creamy-ochre, darker in the centre; bell-shaped; the margin when fresh has a frill of white fibres. Gills black, adnate. The stem, too, as it dries, changes colour from dark brown to cream at the apex. It occurs in rich grassland. Specific name, *sphinctrinus*, means 'banded'.

### *Panaeolus semiovatus*

This species can vary considerably in size from small to medium. Cap averages 4cm; pale creamy-buff; the shape of half an oval as its name implies. Gills black and unevenly spotted; adnate. Flesh pale and slight. Stem is the same colour as the cap, with a black fibrous ring half way down; tall and slim. Grows in horse dung. Specific name, *semiovatus*, means 'half an oval'. *Stropharia semiglobata* is similar.

### *Gomphidius roseus*

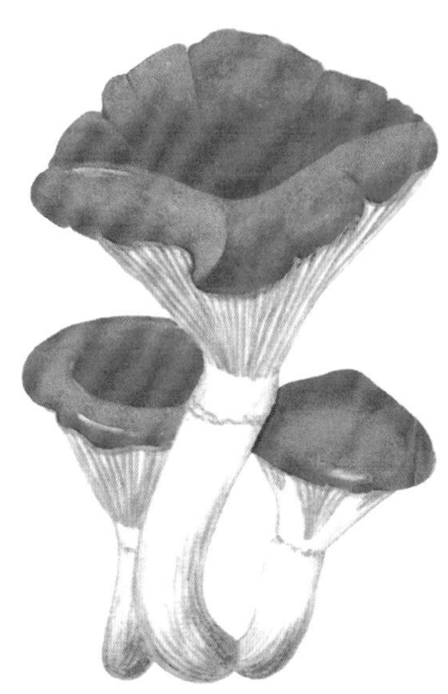

Small to medium-sized species. Cap averages 4cm; pink; slimy; becoming almost flat. Gills white or pale grey; well spaced, forked and running down the stem. Spores blackish. Stem short and sturdy, white. It occurs with conifers, especially in grass under pine. Specific name, *roseus*, means 'rose-pink'.

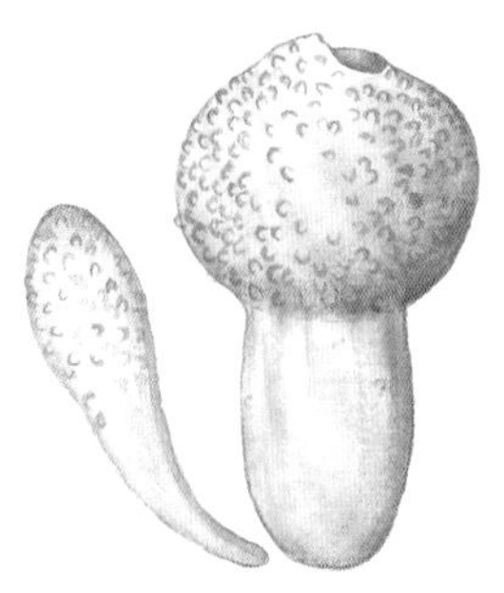

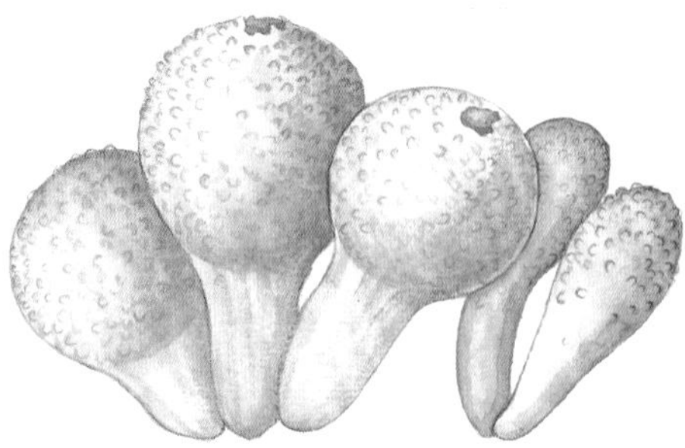

## *Lycoperdon perlatum*

Club-shaped fungus. Size 2–3cm across and 5–10cm tall. Initially pure white but becoming yellowish-brown with age. Its surface is covered with short pyramidal spines, each surrounded by a ring of smaller spines or warts giving it a netted appearance. These soon become loose and rub off, leaving a characteristic pattern. The central spore mass is at first white but it eventually turns grey-black. With age a small mouth opens at the apex from which spores puff out when it is squeezed. Edible when spores are white. Grows in woods. Common name: Puffball. Specific name, *perlatum*, means 'widespread'.

## *Lycoperdon pyriforme*

Small, pear-shaped puffball. Size very variable, but averages 2–3cm across and is not usually much taller. Surface is covered with very small spines that soon fall off, leaving it smooth. The dark spores are discharged through a small central opening. Grows on rotting wood. Specific name, *pyriforme*, means 'pear-shaped'.

## *Langermannia gigantea*

Enormous puffball fungus. It is like a large, creamy-white, smooth football. Eventually it darkens and splits, and is kicked around by livestock, causing the brown spores to be puffed into the air in clouds. Edible when young and the interior is white; it can be cut into slices and fried. Grows in grazed meadows. Uncommon, but seen more often than its frequency would indicate because of its size. Common name: Giant Puffball. Specific name, *gigantea*, means 'giant'.

## *Bovista plumbea*

The size of a golfball or smaller. This little fungus is smooth and initially creamy-white but very soon darkens to the colour of lead. The outer skin is thin but tough and once the spores have formed and dried it is extremely light. At this stage it becomes detached from the ground and blows along the surface in the wind, the spores being blown out of an apical opening. Grows in short grass. Specific name, *plumbea*, means 'lead coloured'.

## *Geastrum triplex*

This is the least uncommon of the Earth Star fungi. Initially formed like a brown onion; the outer layer then splits into 4–7 segments which fold back, leaving a central globe full of spores with an opening at its apex for their discharge. The three-layered outer skin cracks as it folds back, often leaving an uncracked central disc like a cup in which the central globe sits. The whole may measure up to 10cm across. Grows in the soil of woods and hedgerows. Specific name, *triplex*, means 'threefold'.

## *Scleroderma citrinum*

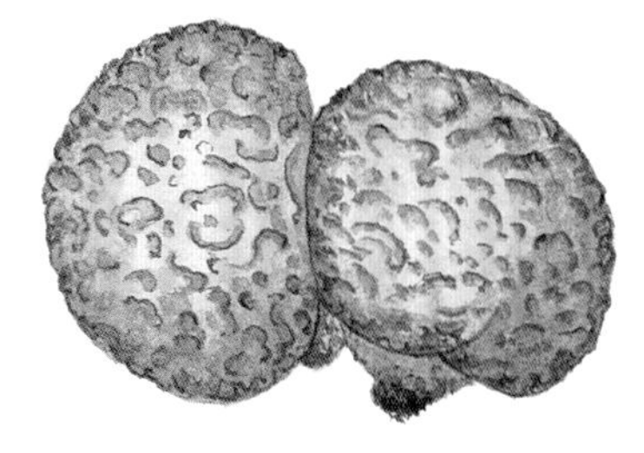

Resembles an old, cracked, tennis ball. It is roughly globose and has a thick skin, dark ochre-brown on the outside, which splits as it expands, showing more and more of the yellow inner layer. The centre comprises a black spore mass. It has no opening for spore discharge, but splits irregularly. Grows in woodland. Common name: Common Earth-ball. Specific name, *citrinum*, means 'lemon coloured'.

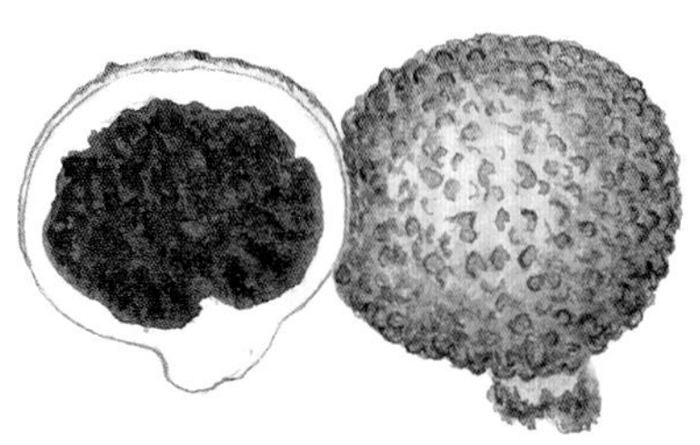

## *Phallus impudicus*

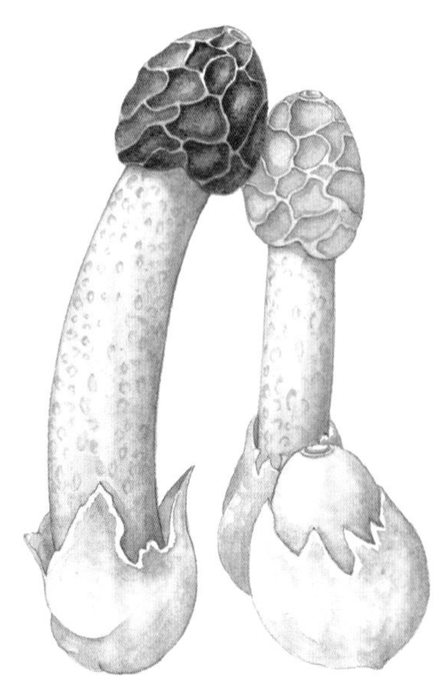

Starts as a soft whitish ball about 4–6cm in diameter and full of a jelly like mass. When mature the case splits and in the course of a few hours the fruitbody rises out of the volva to a height of about 15cm. It consists of a fragile, white phallus-shaped structure, capped by an oval mass of brownish-green mucus containing the spores, and has a characteristic sickly smell. This attracts flies, which rapidly remove the mucus and so distribute the spores. Common name: Stinkhorn. Specific name, *impudicus*, means 'shameless'.

## *Mutinus caninus*

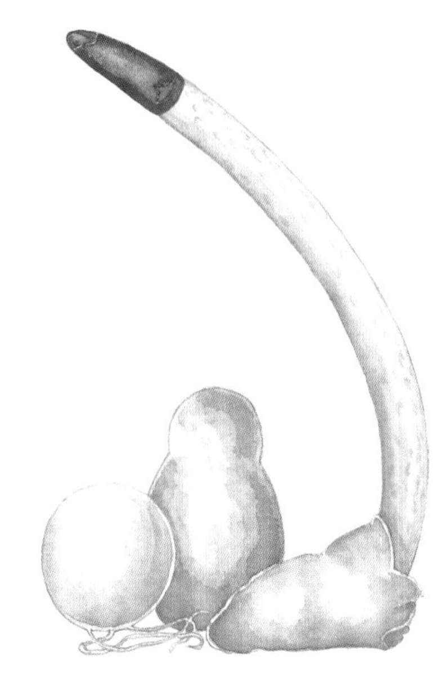

This fungus is similar in its growth to *Phallus impudicus*. The initial 'egg' is about 2cm in diameter and the fruitbody rises to 6–10cm. The stem has a slightly orange tint, and once the gluten is removed the tip is quite a bright orange. Grows in woodland; excavation near a fruitbody will probably reveal other 'eggs'. Common name: Dog Stinkhorn. Specific name, *caninus*, means 'of dogs'.

## *Calocera viscosa*

Branched, antler-like fungus. Size varies greatly from a small growth only 2cm high, branched once or twice, to a tree-like structure 10cm or more in height and branched several times. Colour bright egg-yellow. Grows on coniferous logs or stumps. Common name: Stag's-horn Fungus. Specific name, *viscosa*, means 'sticky'. A smaller, unbranched, yellow relative growing on deciduous wood, especially beech, is *Calocera cornea.*

## *Tremella mesenterica*

Conspicuous jelly-fungus. Forms a convoluted, brain-like mass up to 10cm long and 3cm wide; bright, clear yellow. Especially common on gorse; also often found on many kinds of wood, mainly on fallen branches. Extends its fruiting season well into the winter. Specific name, *mesenterica*, means 'intestine-like'.

## *Exidia glandulosa*

Similar in shape to *Tremella mesenterica*. This is an all-black jelly, forming brain-like lumps from 3–5cm. Its shape differentiates it from the black jelly-like Ascomycete, *Bulgaria inquinans*, as does the fact that a finger wiped across its surface does not become black with shed spores. Grows on deciduous wood. Common name: Witches' Butter. Specific name, *glandulosa*, means 'full of glands'.

## *Auricularia auricula-judae*

Cup or ear-shaped fungus. Average size 3–8cm across. The hollow faces downwards and the back often shows branched wrinkles resembling veins. It is a felty tan colour on the upper surface and a smooth grey-brown below. With the light from behind it appears translucent. Commonest on elder trunks. Fruits right through the winter. Common name: Jew's Ear. Specific name, *auricula-judae*, also means 'Jew's ear'.

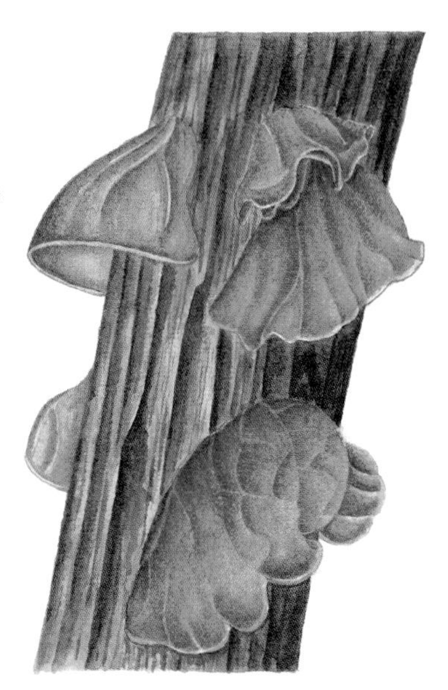

## *Phlebia radiata*

Grows in flat sheets, to any diameter. The growing edge is bright orange-red, fading to greyish in the centre; the surface develops wrinkles at right angles to the edge. Found on the surface of dead wood; if it encounters moss in its growth it will form around it, so destroying the radiating pattern. Dull coloured specimens also occur. Specific name, *radiata*, means 'radiating'.

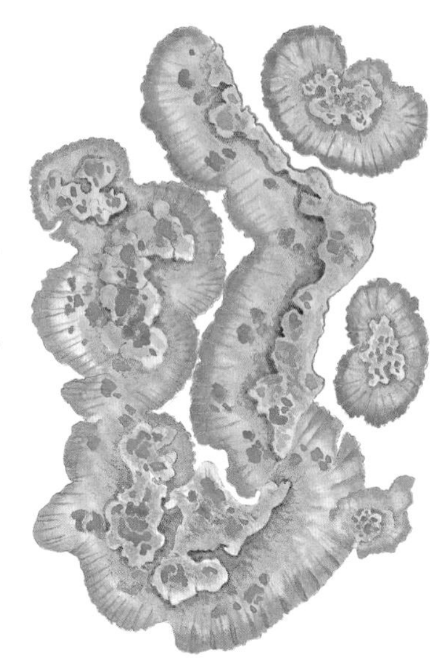

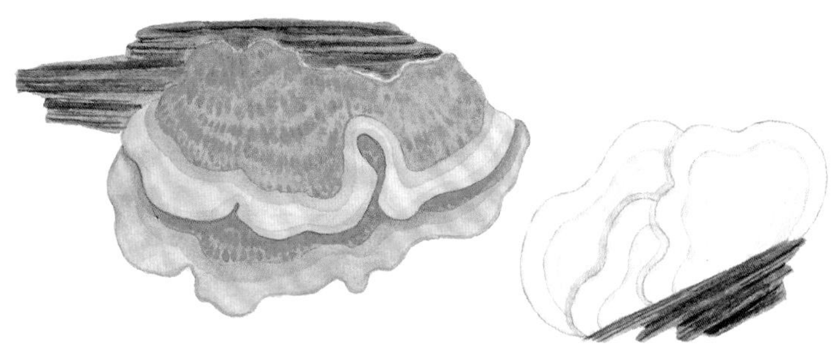

## *Merulius tremellosus*

A gelatinous bracket species. The whole fungus may measure up to 20cm across, though it is more usually half that size. Colour is varying shades of orange, being brighter on the undersurface, which is covered in folds linking together to form pores. The upper surface of the bracket form is covered in white hairs. Grows flat at first on the surface of deciduous wood, then bends outwards to form a shelf. Specific name, *tremellosus*, means 'trembling'.

## *Stereum hirsutum*

This is a tough, flexible bracket. Measures about 2cm deep and 1mm thick. Colour varying shades of yellow on the underside, with the upper surface often zoned in shades of ochre and grey, and covered in fine hairs. Below the bracket itself there may be a variable layer applied flat to the surface of the deciduous wood on which it lives. If the undersurface is bruised by rubbing hard it remains unchanged. Specific name, *hirsutum*, means 'hairy'. *S. rugosum* is a similar species but reddens or 'bleeds' when bruised. *S. sanguinolentum* grows on conifer wood and reddens quite markedly when bruised.

## *Coriolus versicolor*

Perhaps the commonest bracket. It is usually about 5–8cm in diameter but only 2mm thick. The upper surface is zoned in many colours and overall may range from buff to black. It is silky when young but this disappears with age. The underside is pale buff and consists of fine pores. Grows on all types of wood. Specific name, *versicolor*, means 'of various colours'.

### *Thelephora terrestris*

This fungus consists of many small fan-shaped fruitbodies. They measure approximately 2–3cm across. Base colour is reddish-brown but darkens considerably with age. The surface is covered in radiating fibres, which overhang the edge as a white margin. As it has no means of supporting itself, it climbs up grasses, heather and in fact anything available. It grows associated with the roots of conifers. Common name: Earth Fan. Specific name, *terrestris*, means 'growing on the ground'.

### *Craterellus cornucopioides*

Trumpet-shaped fungus. May reach 10cm in height. Horn interior is black when moist; exterior, which sheds the spores, is grey. Edible and well-flavoured. Grows among the fallen leaves of deciduous trees, usually in groups. Common names: Horn of Plenty; Trumpet of Death (despite its edibility). Specific name, *cornucopioides*, means 'resembling a horn of plenty'.

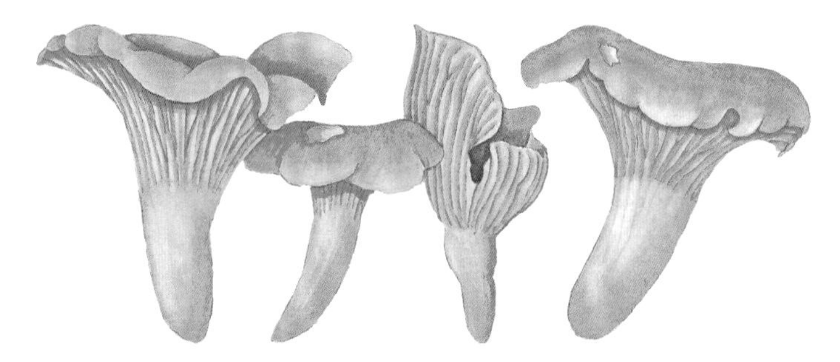

## *Cantharellus cibarius*

Medium-sized fungus. Cap averages 7cm; bright egg-yellow; tending to become funnel-shaped after starting rounded. Appears to have gills but these are in fact branching corrugations in the spore-bearing surface, which is slightly paler than the upper surface. Stem short, sturdy, paler than cap. Smell pleasant, of dried apricots. Edible and of excellent flavour. Occurs in all types of woodland. Common name: Chanterelle. Specific name, *cibarius*, means 'pertaining to food'. *C. infundibuliformis* is the same shape but has a brown cap, and yellow gills and stem. It, too, is edible.

## *Hydnum repandum*

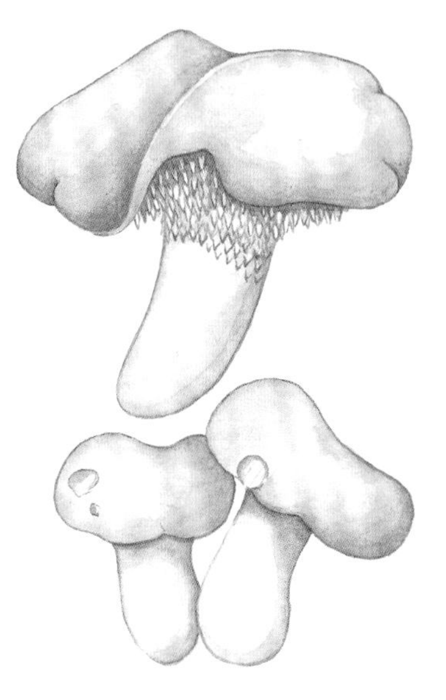

Medium-sized fungus. Cap averages 5cm; pale buff or creamy coloured. No gills; but below the cap there are many spines from 4–8mm long, slightly more pinkish than the cap; these are longest halfway in from the edge and run down the stem, becoming shorter. The spines are brittle and rub off. Flesh has a cheesy consistency and is edible. Grows in deciduous woodland. Common name: Wood Hedgehog. Specific name, *repandum*, means 'bent backwards'. *H. rufescens* is similar but with a slightly rufous cap; the spines do not extend onto the stem.

## *Clavulinopsis helvola*

This fungus consists of a group of pale yellow or orange-yellow 'stalks' with blunt tips about 5cm high and 2mm thick. Grows in short grass. Specific name, *helvola*, means 'pale yellow'. *C. corniculata* grows in similar situations but is slightly darker in colour and is branched. *C. fusiformis* is similar to *C. helvola* but prefers more heathy situations and has pointed tips.

## *Ramaria stricta*

A tree-like fungus. Up to 10cm high, with multiple branches ascending vertically and of a pale cinnamon colour, paling at the tips. It has quite a sweet smell but a sharp or peppery taste. Grows on fallen branches or stumps of both deciduous and coniferous trees. Specific name, *stricta*, means 'rigid'. There are several other variously-coloured species of *Ramaria*, including the similarly-shaped *R. ochraceo-virens*, which differs in turning greenish with age, and on bruising. Found on conifer needles.

## *Sparassis crispa*

Quite a large fungus. Consists of a much-branched and curled pale ochraceous-grey globe, about 15cm in diameter, in shape rather like a large specimen of one of the curly lettuces. It is edible and pleasant, having a taste of anise; but it is very difficult to remove all the sand that collects between its lobes. Common name: Cauliflower Fungus. Specific name, *crispa*, means 'curled'.

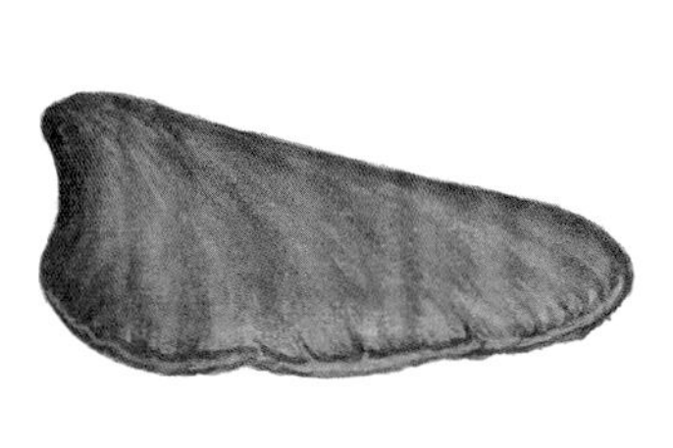

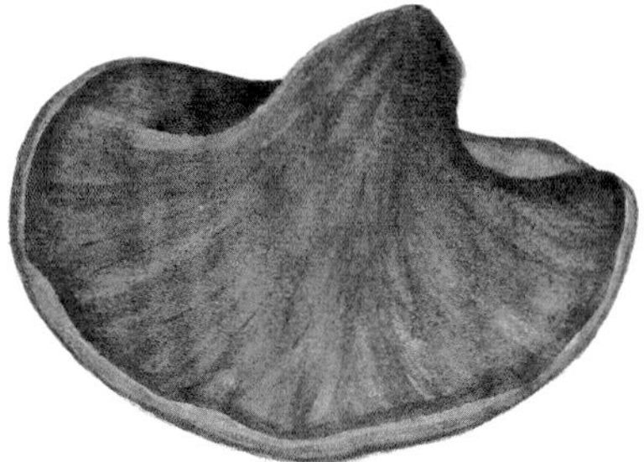

## *Fistulina hepatica*

Large, rather rubbery bracket. Average size 12cm in diameter. When young it is a deep reddish colour, rather like uncooked liver, but later becomes more rigid and darkens. In the young state it oozes drops of blood-like fluid. Grows on oak trees; the mycelium does not destroy the wood but colours it to make the desirable 'brown oak' of cabinet-makers. Common name: Poor Man's Beefsteak. Specific name, *hepatica*, means 'like liver'.

## *Laetiporus sulphureus*

Conspicuous tiered bracket. The individual brackets usually have a wavy edge and are up to 30cm wide; sulphur yellow on the upper surface and more orange on the underside. Edible when young, it is pleasant cooked. Grows on oak and yew in layers, from two or three to a dozen or more, one above the other. Common name: Chicken of the Woods. Specific name, *sulphureus*, means 'sulphur-coloured'.

## *Meripilus giganteus*

One of the largest polypores. It forms a rosette of fan-shaped caps each one 20–30cm across and 2cm thick; felty-brown and zoned on the upper surface. The pores are yellowish-grey, bruising darker. The whole fruitbody may measure almost a metre across in large specimens. Grows on the dead roots of trees, especially beech, appearing on the ground often at some distance from the trunk or stump. Specific name, *giganteus*, means 'giant'.

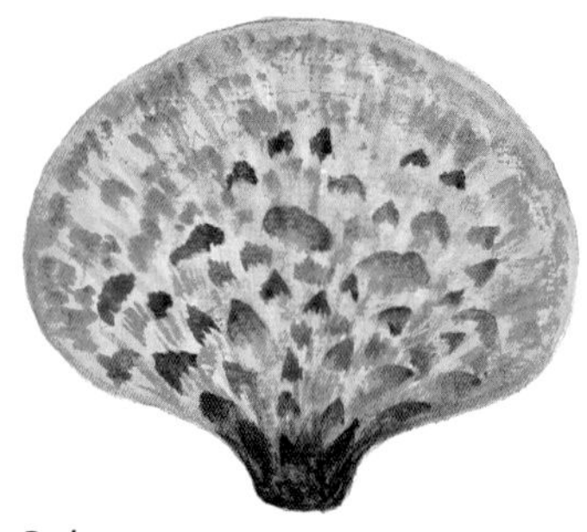

## *Polyporus squamosus*

Very large polypore that grows as a bracket. It forms a semicircular disc up to 50cm across and 5cm thick at the base, coloured cream with large dark brown scales on the upper surface. In texture it is corky-hard when mature. The creamy pores are large, angular and irregular. Grows on deciduous trees, especially sycamore and elm, and sometimes beech as well. Common name: Dryad's Saddle. Specific name, *squamosus*, means 'scaly'.

## *Piptoporus betulinus*

Large bracket. Rounded semicircular outline commonly about 20cm across and 5cm thick at the centre; smooth, pale ochre-brown upper surface; the fine pores are at first creamy-white. Grows on birch trees. Common name: Razor-strop Fungus, as in the past it was dried and used for sharpening razors. Specific name, *betulinus*, means 'of birch'.

## *Pseudotrametes gibbosa*

Thick bracket fungus, of corky texture. Measures up to 20cm wide, 15cm deep and 8cm thick. Upper surface white; the pores are radially elongated, creamy. It can be recognised with a fair degree of certainty by the fact that algae rapidly colonise the upper surface, turning it green. It is rapidly attacked by insect larvae. Grows on dead deciduous wood, mainly beech. Specific name, *gibbosa*, means 'humped'.

## *Ganoderma australe*

Very large and impressive bracket fungus. It may reach almost 1 metre across, and is commonly up to half that size; blackish-brown in colour, wrinkled and grooved in concentric zones. It is perennial, layer upon layer forming below the original bracket and, as it is of woody consistency, its weight is considerable. The pores are fine and pale, turning brown when scratched. In winter, the cocoa-coloured spores form a brown dust around and on the fungus. Grows on beech trees. Specific name, *australe*, means 'southern'. *G. applanatum* is a similar species of more northern distribution and can be identified with certainty only by its spores.

## *Daedelopsis confragosa*

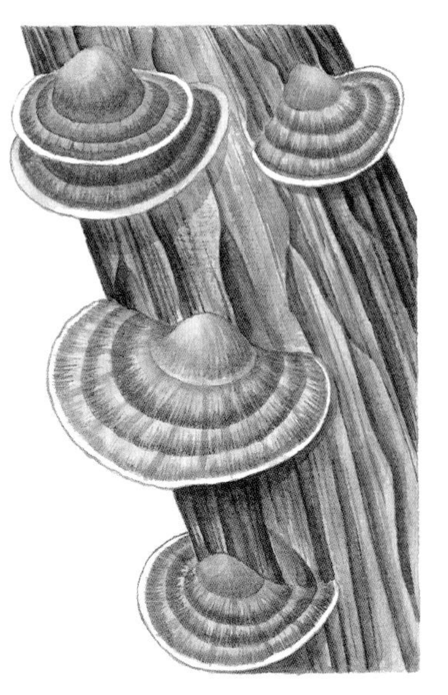

Sharp-edged bracket fungus. Measures up to 20cm across, 10cm deep and thickening to only 2cm. The upper surface is concentrically zoned, initially buff to cinnamon brown, then darkening and at length becoming rusty-red. The pores are at first creamy-buff and bruise pink when fresh, later becoming grey. It may occur on most deciduous trees but is most common on sallow. Common name: The Blushing Bracket. Specific name, *confragosa*, means 'rough'.

## *Fomes fomentarius*

Large, distinctive, hoof-shaped bracket. Measures up to 15cm in diameter. The surface is grey, ridged and zoned with paler brown. Pores light grey-brown. This perennial fungus is very hard and in the past was used as tinder. Extremely common on birches in the north, especially in the Highlands of Scotland; it occurs more rarely on beech in southern England and Continental Europe. Common name: Tinder Fungus. Specific name is derived from *fomentum*, meaning 'tinder'.

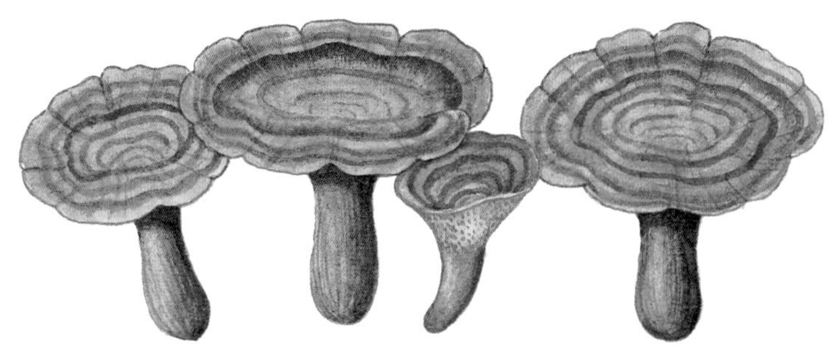

## *Coltrichia perennis*

A centrally-stalked fungus. Cap up to 8cm; the upper surface is concentrically zoned in many shades of brown; the underside has cinnamon-brown pores; thin and dry. Stem usually short. Adjacent fruitbodies may be fused so that it appears to have multiple stems. Grows on the ground, usually on sandy, heathy, acid soils. Specific name, *perennis*, means 'perennial', despite the fact that it is an annual.

## *Aleuria aurantia*

The brightest of the more common cup-fungi. It forms cups of various sizes from 1cm, to 10 or 15cm in large specimens. Inner surface bright orange, outer surface orange-grey; resembles orange peel turned inside out. Sometimes when collected it will give off a puff of spores like smoke as all the asci discharge simultaneously. Grows on the ground amongst grass or leaves. Common name: Orange Peel Fungus. Specific name, *aurantia*, means 'golden'. The similar *Sarcoscypha coccinea* has a more scarlet inner surface, a paler exterior and grows in moss on rotten wood.

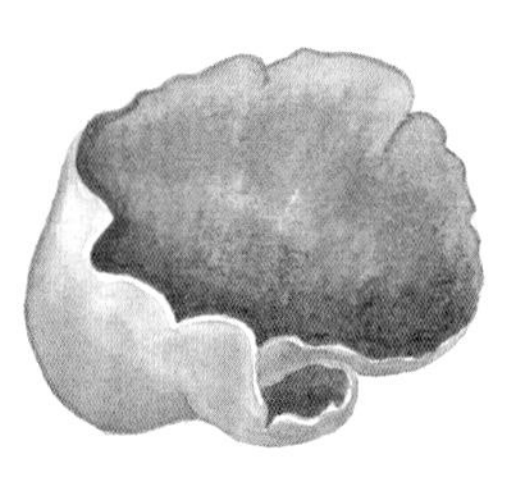

## *Peziza repanda*

Cup-shaped fungus. Usually about 6cm in diameter but can grow to 12cm on occasion. The inner surface is pale brownish-buff and the exterior slightly paler. The edge becomes wavy in larger specimens and is usually irregularly toothed. It grows on rotting vegetable matter and is the only common Peziza also to be found on rotting wood and sawdust (when it may reach a large size). Specific name, *repanda*, means 'bent backwards'.

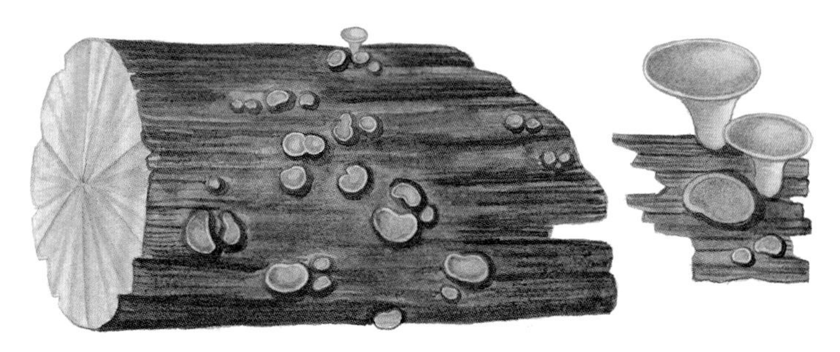

## *Chlorosplenium aeruginascens*

Tiny, bright fungus. The fruitbodies are flattened or cup-shaped, about 1–5mm across; bright verdigris green. Grows on oak and other hardwood logs. The mycelium of this fungus stains the wood green and is more often seen than the fruitbodies. (The coloured wood was formerly used in the decorated boxes known as 'Tunbridge Ware'.) Specific name, *aeruginascens*, means 'becoming green'.

## *Bulgaria inquinans*

Small, jellylike fungus. Each black 'jelly-blob' consists of a thick circular body, averaging 1cm in diameter, which becomes a rounded top-shape, brownish at first on the sides and black on the upper surface. Grows in rows along the bark of recently-fallen oak, and irregularly on the bark of beech. It can be differentiated from *Exidia glatululosa* by its black spores, which come off on the finger, whereas the latter has clear spores. Specific name, *inquinans*, means 'becoming unequal'.

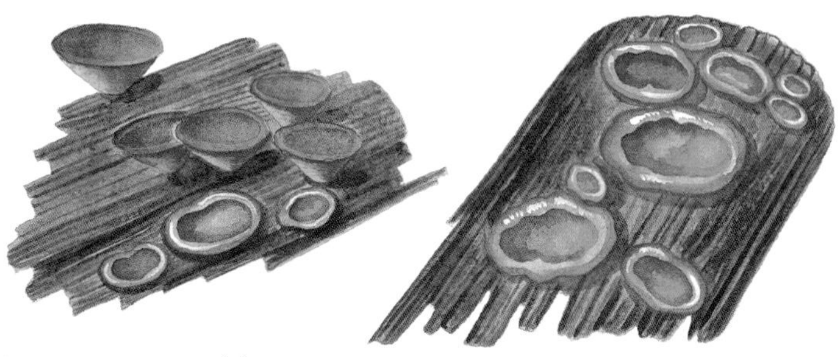

## *Ascocoryne sarcoides*

Jelly-like fungus. Consists of reddish-purple 'jelly-blobs' clustered together and recognisable by their colour. Not all become mature or reach the 'perfect' stage but when they do they become top-shaped, about 1.5–2cm across, with a flat upper surface. Grows on the bark, and sometimes the wood, of fallen beech. Specific name, *sarcoides*, means 'resembling a blotch'. The related *A. cylichnium* is almost indistinguishable in the field.

## *Leotia lubrica*

Drumstick-shaped, jelly fungus. Grows 2–4cm high; dull yellow, the head slightly more olive than the stem. Usually occurs in small groups on the ground, mainly associated with deciduous trees. Not uncommon, but easily overlooked. Common name: Jelly Babies, due to the undoubted resemblance to the old-fashioned sweets. Specific name, *lubrica*, means 'slippery'.

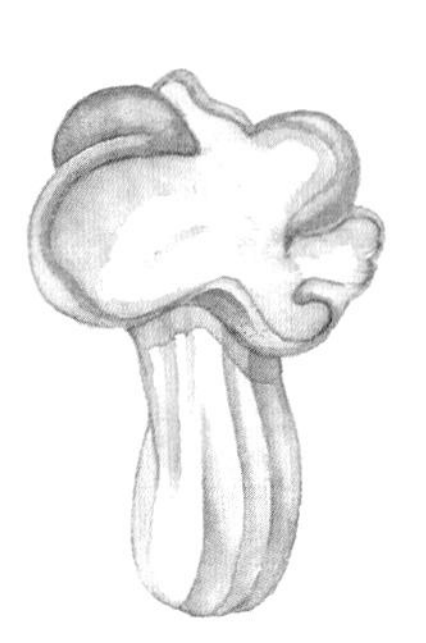

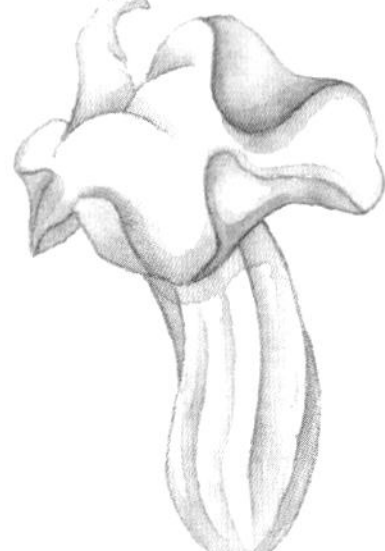

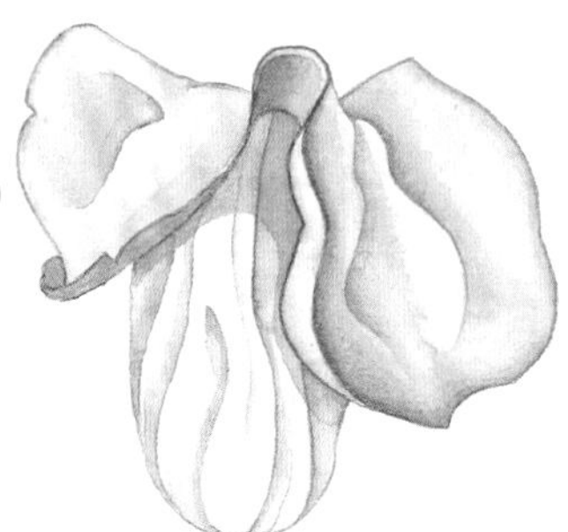

## *Helvella crispa*

The distinctive shape of this fungus makes it easily recognised. Cap may reach 12cm or more in large specimens, but is more usually about 8cm high; very pale grey to whitish; curly and may vary greatly in shape, but always sits on a deeply-furrowed white stem. Occurs in deciduous woods in grass or leaf litter. Specific name, *crispa*, means 'crisped' or 'curled'. *H. lacunosa* has a dark grey to black cap and a grey stem.

## *Morchella esculenta* **POISONOUS** ☠

Distinctive fungus, highly regarded as a table delicacy. Extremely variable in size, ranging from 7–30cm high. Head pale fawn, wrinkled. Stem white and thick. Found in a variety of habitats from sand dunes to chalk woodland, but appears to prefer alkaline soils. Occurs in the spring. Uncommon. Common name: Morel. Specific name, *esculenta*, means 'edible'. *Gyromytra esculenta* is similar in appearance but has reddish-brown and more rounded convolutions. Poisonous. Grows with conifers. Also occurs in spring. Common name: False Morel.

## *Hypoxylon fragiforme*

Tiny fungus. Appears as small spheres averaging 10–15mm, rusty-red at first, then becoming black. The surface is minutely rough. As the red phase is short lived and tends to occur in late summer, the fungus is usually encountered later in the season in the black state, in which it is very persistent. Grows on the bark of dead beech. Specific name, *fragiforme*, means 'appearing broken'.

## *Daldinia concentrica*

This fungus forms in successive layers. Appears as round, shiny black lumps about 5cm in diameter. First it is covered with a reddish spore layer but this washes off, leaving the black surface. If it is cut in half it will be seen to be composed of concentric layers of very dark to very light grey material. Grows on ash; rarely on other deciduous trees. Common names: Cramp Balls (because of its old reputation for curing cramp), or King Alfred's Cakes. Specific name, *concentrica*, means 'concentric'.

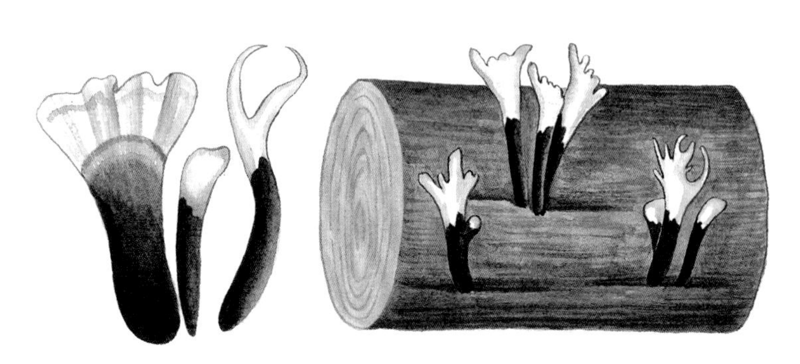

## *Xylaria hypoxylon*

This is one of the most common fungi on dead wood of all kinds. It consists of flattened, black stems, about 3cm tall, arising vertically from the wood and usually (but not always) branches into two- to five-pointed antler-like extensions. At first these are white and then as they mature they blacken so eventually the whole fungus is black. Common name: Candle-snuff. Specific name, *hypoxylon*, means 'under wood'.

# Image gallery

▲ *Helvella crispa*, p.96

▲ *Lactarius torminosus*, p.28 ▼ *Lactarius rufus*, p.30

▲ *Russola claroflava*, p.32

▼ *Russula sardonia*, p.35

▲ *Hygrocybe psittacina*, p.39 ▼ *Hygrocybe conica*, p.40

▲ *Amanita muscaria*, p.43

▼ *Mycena galericulata*, p.58

▲ *Lycoperdon perlatum*, p.77 ▼ *Geastrum triplex*, p.79

▲ *Tremella mesenterica*, p.81 ▼ *Coriolis versicolor*, p.84

▲ *Ramaria stricta*, p.87

▼ *Fistulina hepatica*, p.88

▲ *Ascocoryne sarcoides*, p.95

▼ *Xylaria hypoxylon*, p.98

## BIBLIOGRAPHY

Bon, M., *Mushrooms and Toadstools of Britain and Northwestern Europe*. Hodder and Stoughton, London, 1987.

Lange, J. and Hora, F.B., *Collins Guide to Mushrooms and Toadstools*. Collins, London, 1963.

Pegler, D., *Mushrooms and Toadstools*. Mitchell Beazley, London, 1987.

Phillips, R.P., *Mushrooms and Other Fungi of Great Britain and Europe*. Pan Books, London, 1981.

The above are inexpensive, well-illustrated guides which do not rely entirely on microscopic features. Three more specialised books, which cover selected groups, including most British species, and detailed microscopic features, are:

Breitenbach, J. and Kranzlin, F., *Fungi of Switzerland* (vol 1, Ascomycetes). Mykologia, Switzerland, 1984.

Breitenbach, J. and Kranzlin, F., *Fungi of Switzerland* (vol 2, Heterobasidiomycetes). Mykologia, Switzerland, 1986.

Breitenbach, J. and Kranzlin, F., *Fungi of Switzerland* (vol 3, Boletes and Agarics 1st Part). Mykologia, Switzerland, 1991.

## USEFUL WEBSITES

The British Mycological Society
www.britmycolsoc.org.uk

The society publishes a newsletter, Mycologist News, which aims to promote an interest in fungi among people of all ages and educational levels; this is issued four times a year and is free to all members, including Associates (mainly amateurs). It also publishes four journals: Fungal Biology; Fungal Ecology; Fungal Biology Reviews; and Field Mycology.

The Wildlife Trusts
www.wildlifetrusts.org

Most of The Wildlife Trusts hold field meetings, known as Fungus Forays, in the autumn and there is usually an expert in attendance who will identify many of the specimens found. Details of your local Wildlife Trust can be found on the website.

# Index

Page numbers in **bold** refer to photographs

# Index